Motorbooks International Ill 

ries

Illustrated

DATSUN/NISSAN SPORTS CAR BUYER'S ★ GUIDE®

John Matras

First published in 1996 by Motorbooks International Publishers & Wholesalers, 729 Prospect Avenue, PO Box 1, Osceola, WI 54020-0001 USA

Library of Congress Cataloging-in-Publication Data
Matras, John.
Illustrated Datsun/Nissan sports car buyer's guide/John Matras.
p. cm. —(Motorbooks International illustrated buyer's guide series)
Includes bibliographical references and index.
ISBN 0-7603-0136-0 (pbk.)
1. Datsun automobile—Purchasing. 2. Nissan automobile—Purchasing. 3. Sports cars—Purchasing. I. Title. II. Series.
TL215.D35M38 1996
629.222'2—dc20 96-9403

On the front cover: First and last. Datsun redefined the low-cost sports car market with its eminently capable 240Z. This fine example belongs to Haji Arata. The 300ZX may be the end of the Z line. Introduced in 1990, the fantastic, high-tech ZX was discontinued in 1996. *David Gooley*

On the back cover: Three different approaches to the sports car market. Datsun's 1600 roadster was an MG B that didn't leak oil, while the 510 sedan became a sporting car almost by accident; Bob Sharp raced the latter with great success in the late 1960s and early 1970s. The 1984 200SX Turbo offered speed and handling at a reasonable cost. *Automobile Quarterly, Nissan*

Printed in the United States of America

Contents

Acknowledgments

My name is on the cover of this book, but the book wouldn't have been possible without the contribution of help from a number of people. I hope not to overlook anyone, but if I have, I'm sure I'll hear about it, as I should. But particularly I must mention Bill Garlin, Manager of Product News of Nissan North America's Product Public Relations department for coming up with some obscure photographs to put the Illustrated in this Buyer's Guide, as well as data and information that I hadn't been able to find elsewhere. And thanks to all, including those in Japan, who helped Bill in this task. Thanks also to Jim Gill at Nissan NA for helping with numbers, and other Nissan employees who have helped over the years with projects that preceded this one and in the long run made this project possible.

Thanks also to Walt Koopman, for his knowledge of Datsun and Nissan cars and for his research assistance and lending research materials; Jonathan Stein, Publishing Director, and John Heilig, former Associate Editor, for sharing their time and the resources of *Automobile Quarterly;* Julianne Purther, Assistant Art Director, *Car and Driver* magazine, for photo assistance; Craig and Elaine Halstead for Datsun Roadster information; Wayne Kinch, of the Twin Turbo Z Association; Michael F. Hollander, author, *The Complete Datsun Guide;* Jenny and Paul Lagowski of the White Rose Z Club; Larry Saavedra, Editor, *Sport Compact Car* magazine; and Jeffrey Bartlett.

Thanks to Mike Dapper and Zack Miller, editors of this tome who otherwise would go unacknowledged and shouldn't.

And thanks to my wife Mary Ann, for her support and for putting up with a writer's weird schedule, and to my daughters Amanda, Carolynn and Katherine for just being themselves.

Introduction

Datsun and sports cars have been linked in America almost since the very beginning, and though other Japanese automakers have built sports cars—most notably Mazda and Toyota—perhaps none has been more dedicated to the genre than Nissan. The manufacturer has had a genuine sports car on the U.S. market from 1959 through 1996, and has sold a wide variety of sporty coupes over the years as well. Some have been better than others, of course, and certainly some have been made to appeal more to the enthusiast market. It is the latter that this book addresses.

No doubt some will criticize omissions as well as some models included here. Should the Maxima—which Nissan for a while called its "four door sports car"—have been covered? Then what about the Z-motored 810? Or since the 510 is included, why not the 610? And why include the original but ugly 200SX? The answer to all the questions is that the decisions were largely subjective. Room simply didn't exist to list everything that Nissan sent to America, and who wants to know everything there is to know about the B210 or the Axxess mini-van? The first 200SX was put in mostly for historical reference, although undoubtedly someone out there has half a dozen stashed in a barn waiting for the inevitable price surge. In the end, a man's gotta do what a man's gotta do.

A man's also gotta, in this series of books, include a five-star investment rating guide, and let it be said from the beginning that no one will get rich buying Datsun futures. Prices simply won't go up that much. On the other hand, the 240Z has begun to receive interest on the collector market and so, to a lesser extent, have the roadsters. The investment rating used applies only to this book, with the highest rating of five stars applied to the best prospect included, i.e., the 240Z. Cars of minimum investment potential are rated one star—although the 1976 200SX was awarded only a half star. The presumption is that if the cars were worthy of being included, they were worthy of some sort of rating. Ratings were also automatically biased toward older cars receiving higher ratings as the newer cars haven't even fully depreciated yet. And as a general rule, the sportier the car—more power, turbos, whatever—the more collectible it will be.

Thus while the early Z cars are beginning to climb in value—on par with, for example, with Volvo 1800s and such—they'll never make you rich, much less put your kid through a semester in the Ivy League. The roadsters are climbing too, and they're relatively scarce as well. But there's a feeling out there that Japanese cars just can't be classics, and that general opinion, alas, will be slow to go away. On the other hand, it means that Datsuns and Nissans will remain relatively affordable. Forget the gilded pedigree. We know a good car when we see one.

Chapter 1

DAT to Now: A Brief History of Nissan

The first Nissan sports car, in the broadest sense of the term, was the 1952 Datsun Sport. It had all the hallmarks of the traditional British sports car—the double-humped dash and a fold-flat windscreen, headlamps on the radiator shell, cut-down "suicide" doors—but all arranged in a semi-comical manner. It was a caricature, perhaps, of the Singer 1500, which was itself a caricature of the MG TD. Bearing the model code DC-3, the Sport used the 20-hp

Ancestor to every Datsun/Nissan sports car, the DAT 41 was produced for almost a decade. *Nissan*

Called first the Datson Type 10, the name was changed in 1932 to Datsun, making it the first model to wear that name. *Nissan*

Datsuns of the Thirties, such as this Type 14 of 1934, looked like shrunken American cars. *Nissan*

860-cc side-valve four-cylinder engine and three-speed crash-box transmission of the DB-2 sedan, yielding a top speed of 43 mph. But no matter how fast it could go, it wasn't what Nissan—or Japan—needed at the time. It may not even have been a Japanese idea, as Nissan had been under American administration following the war. Regardless, it was for better or worse a dead end.

But it had hardly been the beginning for Nissan, which can trace its ancestry to the 1911 founding of the rather presumptuously named Kwaishinsha Motor Car Works. Accounts differ, but the consensus has Masujiro Hashimoto, an American-trained engineer, as the founder. His first automobile was produced in 1912 but was less than satisfactory. The second moved under its own power in 1913 and by 1914 was in series—if very limited—production. The name of the new car, DAT, has been attributed to the initials of three surnames: Kenjiro Den (who had helped establish the company), Rokuro Aoyama (Hashimoto's childhood friend), and Meitaro Takeuchi (a benefactor). No doubt to suggest lively performance, the acronym also spelled the Japanese word for "hare."

The first model was replaced in 1916 by the DAT 41, with a conventional front engine-rear drive layout, and built with open and closed bodywork. Its 1.5-liter L-head four produced 15.8 hp, and it had a multi-disc clutch, a four-speed transmission, and hypoid bevel final drive.

The DAT 41 remained in production for more than a decade. With few domestic competitors, there was little need for change. By 1923, only 12,700 automobiles were in service in Japan, and despite horrendous import duties, fewer than 1,000 had been built in the Land of the Rising Sun. Car ownership was limited to the wealthy, and the use of cars was restricted to urban areas since Japan's country roads were worse even than those in the turn-of-the-century United States. Japan's rural bridges were so narrow that cars wider than 40 inches couldn't pass. Trucks for industry and the Imperial army were a much more significant market and more important to Kwaishinsha's economy.

Financial repercussions of the great earthquake of 1923, however, threw Kwaishinsha together with Jitsuyo Jidosha Seizo (figuratively speaking). The latter company had been formed for aircraft production but, when that failed, turned to motorized tricycles and, around 1920, the four-wheeled Lila, a lightweight automobile that quickly proved popular as a taxi. Around 1926, the two companies formalized their financial relationship, merging to form the DAT Jidosha Seizo Co., and began production of industrial vehicles. A return to car making was reflected in subsequent name changes, first to DAT Motor Car Co. and then DAT Automobile Manufacturing Co., with the 495 cc DAT 91 appearing in 1930.

DAT became part of the Tobata Imono Co. in 1931, and when a new car, the Type 10, was introduced that year, it was called the son of

The Datsun 15 was launched in 1935. *Nissan*

The 1937 Datsun 70 was an American Graham-Paige built in Japan, the first Japanese big car. *Nissan*

DAT or "Datson." That, however, sounded too much like the Japanese word for "ruin," and when the company's new factory in Jidosha was destroyed by a typhoon, only a modicum of superstition was needed to prompt a name change, especially since customers weren't crazy about driving a car named Ruin. The trade name became Datsun in 1932. This new marque was either, your choice of reference, an "invocation to the sun for protection against further disaster," or simply "deference to the ancient Japanese sun symbol."

In 1933 the Tobata Imono group became Jidosha Seizo Company Ltd., of Yokohama, and a year later was renamed Nissan Motor Company Ltd. "Nissan" was a contraction of Nippon Sankyo, or "Japanese Manufacturing Company." Automobile production began in earnest, with exports beginning in 1935.

The model lineup saw the Datsun 91 joined by the Nissan 70, a car that looked very much like the American Graham Six. In fact, Graham-Paige had sold the tooling and related equipment to Nissan, enabling it to build the first Japanese big car. Production was limited, however, by the small number of people able to afford such a large car. Production of trucks, however, soared as war approached, with load capacities ranging from 1/2 ton to 12 tons.

As in the United States, car building in Japan stopped during the war. After the war, under American administration, truck production was emphasized to aid Japan's reconstruction, there being virtually no demand for private cars anyway. Nissan's first passenger cars, assembled in the Yoshiwara factory where engines were made during the war, came off the line in 1947, essentially an 850cc version of a prewar design.

Like most other Japanese car makers after the war, Nissan entered a license agreement to build foreign-designed cars in Japan. Isuzu made Hillmans, Hino built Renaults, and Mitsubishi assembled Kaiser-Fraiser Jeeps. A 1952 agreement allowed Nissan's production of a version of the Austin A40, followed in 1955 by an A50 variant. Content was almost exclusively of Japanese origin by 1956, but the English influence in technology would continue, particularly the 850 cc side-valve engines.

The 1951 Datsun DS-2 Thrift was a crudely styled version of the company's prewar models. *Nissan*

The short-lived Datsun Roadster of 1952 was inspired by the British Austin 7. Japan lacked the wealth to support a sports car. *Automobile Quarterly Photo and Research Library*

Nissan's car production climbed relentlessly during the '50s, from only 865 in 1950, to 7,800 in 1955, and 66,000 in 1960, keeping Nissan in the lead in Japan. The company's first American sortie was to the 1958 Imported Car Show in Los Angeles, where the 210—called the "1000" for American consumption—was displayed, along with Toyota's Toyopet Crown, as two new models for the U.S. market. The theory for both companies was that a big country that bought a million cars per year could easily buy 5,000 Japanese cars. But only hundreds were sold. The cars were simply less than suited for American conditions. Designed for Japan's 35 mph average speed, they were unable to keep up with American freeway traffic. Toyota was reduced to selling Land Cruisers, but Nissan's new 310 Bluebird sedan adapted better to America.

The 1.2-liter Bluebird took 24.3 seconds (sec) to reach 60 mph and could cruise at 65

The Type 110 was Datsun's first modern car. The visually similar Type 210, also known as the 1000 Sedan, was introduced to America at the 1958 Los Angeles Imported Car Show. *Nissan*

The Datsun 410 was introduced in 1963. *Nissan*

mph, but *Motor Trend* noted that its taxi background suited it better for "traffic, parking and short trips," and noted that it "should have a long life where heavier-than-average loads for a small car are hauled over roads that are unimproved or in bad shape." It was sold in the United States alongside the Fairlady sportster and the Patrol 4x4.

The Datsun 410 was introduced in 1963 as the 310's successor and was better than its predecessor in every way. Grabbing more attention, however, was the Datsun SPL-310 1500 Sports, particularly when it was revised and expanded as the SPL-311 1600 Sports. Thanks to competition, the sports cars garnered far more publicity than their sales numbers would suggest. Nissan Motors Corp. in U.S.A., the American subsidiary formed in 1960, established a competition department in 1967 under the leadership of West Coast Sports Car Club of America (SCCA) racer Dick Roberts. A Porsche-turned-Datsun 1600 driver, Roberts knew what amateur racers needed. Although the performance potential of Datsuns had already attracted racers—even a first generation Fairlady had been raced in California—the company quickly carved out its space in American road racing.

Among those who campaigned Datsuns with repeated success for many years were Connecticut Datsun dealer Bob Sharp (racing on the East Coast) and, out West, former Shelby employee Pete Brock. Sharp scored first, winning an F-production SCCA national championship in 1967. When the 240Z was introduced, John Morton took two C-Production championships in a row in BRE (Brock Racing Enterprises) Z-cars. Sharp won C-Production in 1972, 1973, and 1975. (Walt Maas, Elliot Forbes-Robinson, Logan Blackburn, and Frank Leary filled in the other years through 1978.) BRE also gave Datsun the SCCA's professional sedan racing championship, the Two-Five Challenge (the Trans Am's junior league), in 1970 and 1971.

The 510 had been introduced in 1967, and though incredibly spare by today's expectations, the car's roominess, durability, and ride comfort set new standards for small cars. The 510 gave Datsun a sedan that was at least the equal of the roadsters. It was Roberts' idea to promote the boxy sedans via racing, and even today the humble economy sedans are prized as vintage racers and IT class competitors. They are even modified, lowered, and hot-rodded for road use.

The Datsun Coupe 1500, also known as the Sylvia, was a coupe version of the 1500 roadster styled by American Albrecht Goertz. *Automobile Quarterly Photo and Research Library*

The 510 was joined in 1970 by the 1200, a smaller car named for its displacement in cubic centimeters. But while Datsun's U.S. sales had risen from almost 22,000 units in 1966 to over 100,000 in 1970, Toyota surpassed Datsun in 1967 as the number one Japanese import, and almost 185,000 Toyotas were sold in the United States in 1970. (Volkswagen led all imports with sales of 569,182 units in 1970.)

The 510 had the virtue of being plain and simple, but to woo more luxury-oriented buyers, Nissan introduced the Datsun 610 for 1963. Powered by a 1.8-liter version of the 510's engine, the 610 had a plush interior and, it could be argued, prettier lines. The 610 was joined by the 510's marketplace successor, the 710, which shared the 610's engine but in a peculiarly styled body with a more basic interior and a simple live axle. Two-liter versions of both were also offered.

The 1200's mini-car replacement in 1974 was the B210 (Datsun Competition actively supported both cars with a great deal of success), another rear-drive chassis. The B210 and 710 sold sufficiently well, as did the 260Z. The fuel crisis of 1974 initially hit Nissan—and everyone else—hard, but sales recovered enough to make Datsun the leading import—over Toyota and even Volkswagen—in 1975.

Nissan entered the front-drive stage with the F-10. This 1976 debutante easily wins the honor of the oddest Datsun ever, with unfortunate styling details assembled in a most haphazard way. The 1977 200-SX, a small sport coupe with a 2.0-liter engine, had some peculiar styling cues of its own. The 1978 model year saw two new models, one a new 510 series, Nissan's vain attempt to recapture the magic of the earlier model. The other was the 810, a luxury model with the six-cylinder Z-car engine.

The Z-car itself became a ZX in 1979 with the debut of the larger, softer, more luxurious 280ZX—which, incidentally, Paul Newman (later to become famous for his salad dressing) drove to an SCCA C-Production national championship the first year out. That same year, an Electramotive-prepared Z-car took the IMSA GTU title for Nissan and driver Bob Devendorf, who also chalked up the 1982 IMSA GTO title.

Sales grew steadily, from 350,403 units in 1976, to 488,217 in 1977, 432,700 in 1978, and 574,166 in 1979. Sales topped 600,000 in 1980, and Nissan was the world's fourth-largest manufacturer of motor vehicles, following only General Motors, Ford, and Toyota. The 310 had replaced the unhappy F-10 as Nissan's

The incredibly homely Datsun F-10, available as a three-door wagon or "sporty hatchback coupe" (shown), was the marque's first front-wheel-drive car sold in the United States. *Automobile Quarterly Photo and Research Library*

front-drive subcompact in 1979, the Sentra arriving in 1983 along with the Pulsar sports coupe. The front-drive Stanza replaced the rear-drive 510 in 1983 as well, and the 810, having adopted the Maxima name in 1980, went to front drive and V-6 power in 1985.

Enthusiasts, though, greeted the V-6-engined 300ZX, arriving in the 1984 model year. It was bigger, better, and stronger, but hardcore sports car fans still missed the purity of the original 240Z.

The winning ways weren't changed, however. One of the nicest guys in racing, the late Jim Fitzgerald, drove a 300ZX to a 1984 SCCA GT-1 national championship, while Morris Clement, in a 280ZX, took the GT-2 title. Newman capture GT-1 in 1985 and 1986, with Scott Sharp, son of Bob, taking the laurels in 1987. A Nissan-backed effort in GTP came to a championship in 1988, Geoff Brabham earning the driver's title. The manufacturer's title, however, went to Porsche, but in 1989 it would be both Brabham and Nissan as champions in GTP, with ten wins in fifteen starts.

Nissan had another winner in 1989. The 200SX, which had gone from ugly duckling to a sporty coupe with turbo power and even a 3.0-liter V-6, made way for the 2.4-liter 240SX, but kept rear-wheel drive. A new front-drive Nissan sports coupe debuted in 1991, replacing the second generation Pulsar. The '91 NX 1600 and NX 2000 shared chassis and engine with the Sentra, so a 2.0-liter hot rod version of Nissan's econobox was issued as the Sentra SE-R. A new 200SX replaced both in 1995. The Stanza became the Altima in 1993 (although Nissan put the Stanza nameplate on it *real small* just so the name might be revived later . . .).

Nissan issued a stunning new 300ZX in 1990, and whether naturally aspirated or twin-turbocharged, the car was an instant hit with critics and a favorite with road testers, although its high price—due in part to high-tech content and in part to the rising yen—has held sales to lower levels than Nissan would have liked. And that, along with federal side-impact rules that would have required an expensive redesign not justified by sales volume, doomed the Z-car sports car at the end of the 1996 model year. Being a technology leader is important but, even as far back as 1911, selling cars and making a profit must always be the bottom line.

The 1978 Datsun 810 sedan, with its inline six-cylinder engine, was billed as the luxury car with the heart of a Z-car. A wagon was also available. *Automobile Quarterly Photo and Research Library*

Chapter 2

1959-1963 Datsun Fair Lady

One legend has it that a Datsun executive was so enthralled with Lerner and Loewe's 1956 musical, *My Fair Lady*, that when the company introduced a sports car version of the Datsun 1000 sedan, it was dubbed the "Fair Lady." Perhaps so, or maybe there's some other reason for the name that, like those of so many Japanese home market models, strikes the American ear as humorous or at least odd.

Regardless, in 1959 Nissan supplemented the Datsun 1000 sedan on the American market with a chubby-cheeked roadster. It was,

Datsun's first sports offering to America had four seats, a fiberglass body and a Healeyesque flourish on the side. *Automobile Quarterly Photo and Research Library*

like many sports cars, based on sedan underpinnings, and as the Datsun sedan looked like an Austin sedan, the new roadster looked akin to certain British roadsters, though one may not be sure just which. Its lines were curved to excess, like a longer car viewed in a fun house mirror. To break up the expanse of the side, a long sweep of trim arced down the side, containing an area of a second paint color. The grille was a simple oval with horizontal slats and a single vertical bar, and the headlamps were set in the leading edge of the fenders. The body had a slight Coke-bottle shape to it, the rear fenders slightly raised and sporting single taillamps.

The body was avant-garde in one regard. It was fiberglass, a fairly novel material for bodywork, but if General Motors used it for the Corvette, it couldn't be inappropriate for Nissan's little sports car.

Neither the flat windshield nor the convertible top with separate frame and fabric were particularly advanced, however. Of course visibility was excellent with the convertible top down, But the top looked peculiar when erected, tall and angular, and it didn't appear as if it would stand much of a breeze. Naturally, side curtains completed the weatherproofing, if it can be called that. The top unsnapped and rolled off and the framework was disassembled; this framework and the spare tire filled the tiny trunk. A tonneau was standard equipment.

The Fair Lady answered a question that perhaps should not have been asked: Where are the four-place sports cars? Coming from Japan.

Unlike a number of roadsters with a padded shelf that was called a seat, the Fair Lady had a genuine back seat. Two of them, in fact, each a 17x17-inch pad with a 17-inch backrest and 36 inches of headroom with the top raised. Knee room was called adequate, per *Motor Trend*, "as long as the front seats are adjusted even slightly forward."

When introduced, the Fair Lady had the same 988-cc pushrod "Model C" four-cylinder "E type" engine that powered the Datsun 1000

Note solid front axle and rear semi-elliptic springs on this home market 1957 SPL211. *Automobile Quarterly Photo and Research Library*

sedan, and with a mere a 37 bhp, the roadster was anything but a rocket. Early in 1960 the Fair Lady was switched to the 1,189-cc "Model E" engine of the Bluebird. It was a sizable increase, raising power to 48 bhp, a 30 percent increase. The engine had solid lifters, a 7.5:1 compression ratio, and a Hitachi/Solex carburetor. It was enough to rate a model designation change. The 1959 Fair Lady was officially the SPL-211 ("S" for "sport," and "L" for "left hand drive"), while the 1960 Fair Lady became the SPL-212.

A lot more power in 1961 warranted another change in model designation, the faster-than-ever Fair Lady called the SPL-213. Power was up to 60 bhp, thanks to a boost in the compression raised to 8.2:1 and a two-barrel Hitachi/Solex carburetor with progressive linkage. The engine earned a new designation, "E-1," in the process. So even though the SPL-211 needed about 45 sec to run the quarter-mile, the SPL-213 fairly knocked it off with 24.3-sec blast.

The Fair Lady was not just a sexy (relatively speaking) body on the sedan's pressed steel frame. Beginning in 1960, the Fair Lady had 14-in wheels with 5.20x14-in tires. The double A-arm front suspension used longitudinal torsion bars (the Bluebird coil springs)

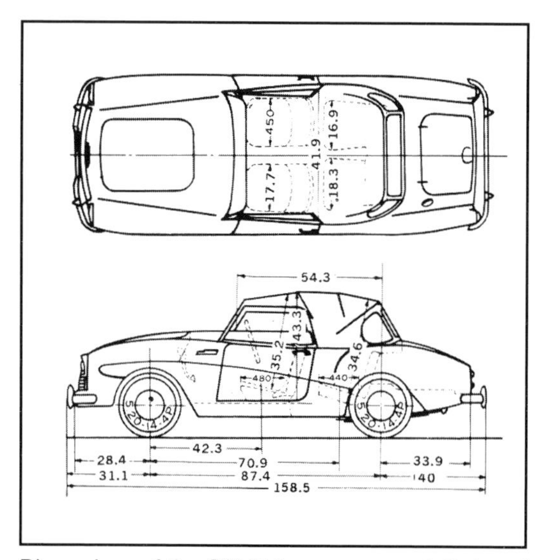

Dimensions of the SPL212 were compact. Outward visibility must have been dreadful with top raised. *Nissan*

along with an anti-roll bar and tube shocks. At the rear the live axle was on leaf springs with a Panhard rod.

The sports car had a floor-shifted four-speed manual (top three synchro), compared to the sedan with only three gears but all synchro. The Fair Lady also got bigger brakes, 10-in diameter by 1.75-in wide (versus the previous size of 8-in diameter by 1.4-in wide), while the sedan had a longer wheelbase and wider track.

And as always, the sports car was more expensive. The 1960 Fair Lady listed for $1,996, versus $1,616 for the Bluebird. The 1961 Fair Lady cost $2,132 against the Bluebird's $1,696. Of course, if the price seems high, remember that a Fair Lady included as standard equipment whitewall tires, "vinyl leather" upholstery, carpet, a dual-tone horn, and a cigarette lighter.

Americans never were particularly enthralled with the Fair Lady. It didn't particularly suit American conditions or American desires for an automobile, and it was truly about 10 years out of date when introduced. Add to that a dealer network that, with 60 stores nationwide, didn't really deserve the name, and Americans weren't falling over each other to be the first to buy. Nissan reports sales of about 123 model SPL-212s and about 106 model SPL-213s, with no data available on the SPL-211. But the Fair Lady, along with the sedans, wagons, pickup trucks, and the Patrol sport-utility vehicle, would help train Nissan in the ways of the U.S. market. Although the Fair Lady name would be discontinued for North America when the SPL-213's successor arrived in 1962 (they were learning already), the chubby little sports car was only the beginning of a lineage that would be the most popular sports car of all time.

What to Look For

Good luck. You could build more Model T Fords from reproduction parts lying about in warehouses than the total number of Fair Ladys (Fair Ladies?) imported by Nissan between 1959 and 1962. It's more likely one will find you first. Surely there's one in a garage or barn somewhere just waiting for you, and more than likely the owner will let you have it cheap, or even free for towing it away. (If not, leave your number and wait. You'll probably get a call in a week or two.)

Restoration will be as difficult and expensive as for any car for which parts are no longer available and for which support organizations are nonexistent. The normal advice to "buy as good as possible" really doesn't hold here either. You're likely to find only one. If you find more, buy all of them. You'll need them for parts.

Please don't expect a restored 1959-1962 Fair Lady to bring a lot on the collector market, either. Despite the roadster's historical significance, there's virtually no demand for a restored Fair Lady. Satisfy yourself with having one of a kind at car shows and having preserved a part of history.

1962 Datsun Fair Lady
Base Price: $2,132
Major options: Transistor radio, $65

General

Layout	rwd
Curb weight (lbs)	1,960
Wheelbase (in)	86.6
Track (front/rear; in)	46.5/46.2
Overall length (in)	158.5
Width (in)	58.9
Height (in)	54.3
Ground clearance (in)	n.a.
Seats	2+2
Trunk space (cu-ft)	"limited"
Fuel tank capacity (gal)	n.a.

Chassis and Suspension

Frame type	Box section ladder-type frame
Front suspension	Double A-arms, longitudinal torsion bars, stabilizer bar, tube shocks
Rear suspension	Live axle, semi-elliptical springs, tube shocks
Steering type	Worm and roller
Steering ratio	n.a.
Tire size (in)	5.20x14
Wheel size/type (in)	14x4/steel disc

Brakes

Front type/size (in)	Drum; 10x1.75
Rear type/size (in)	Drum; 10x1.75

Engine

Type	OHV Inline 4-cylinder
Bore & stroke (mm)	73x7
Displacement (cc)	1,189
Compression ratio	7.5:1
BHP @ rpm	60 @ 5,000
Torque @ rpm (ft-lb)	67.3 @ 3,600
Fuel supply	Two-barrel Hitachi/Solex carburetor
Fuel required	Regular

Drivetrain

Transmission speeds/types	4/manual
Gear ratios (:1)	
1st	n.a.
2nd	n.a.
3rd	n.a.
4th	n.a.
5th	n.a.
Final drive ratio	4.625:1

Performance (Source: *Motor Trend* 12/61)

0-60 mph	24.3 sec
1/4-mile	24.3 sec @ 60
Top speed	n.a. (Mfr.'s claim: 82 mph)
Fuel economy (mpg)	23-26 mpg (observed)

Chapter 3

1962-1970 Datsun 1500 and 1600 Sports Roadsters

★★★	**1962-1966 1500 Sports Roadster (SPL-310)**
★★★1/2	**1965-1970 1600 Sports Roadster (SPL-311)**

Change sometimes goes unnoticed. We often miss the turning of the tide in the roll of the waves. So perhaps it's not surprising that *Road & Track*'s review of the 1962 New York Auto Show didn't include a mention of the new Datsun 1500 Sports. Fair enough in that Japanese cars had not yet made much impact on the American market. How much credence was there in the idea of a genuine sports car from a country still better noted for trinkets than technology?

But this Japanese sports car would soon be compared favorably to venerated European marques and, with more power and refinement, would change from "a good car for your wife," to one of the fastest sports cars in the affordable price range. It became a mainstay of SCCA production-class racing, and better-not-older Datsun roadsters compete even to this day. It left production only in deference to the 240Z.

It's significant that Datsun chose New York for the 1500's international debut—after showing a prototype at Tokyo the preceding fall—as the model was aimed at the U.S. market. Although the roadster is often called a copy of the MGB, it would not seem to be so unless MG had been outwitted by Datsun's own secret agents. The New York Auto Show

The Datsun 1500 was a real breakthrough, Nissan's first genuine sports car. Note side molding from headlight bucket to tail. *Nissan*

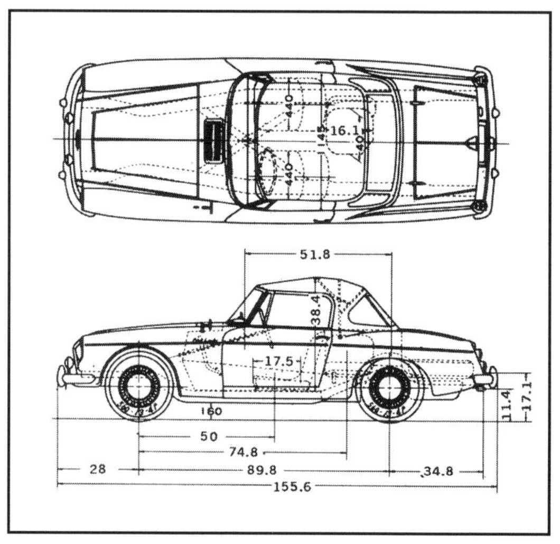

The dimensions of SPL310 1500 were spot on for a traditional English sports car—but it came from Japan. *Nissan*

DRIVE A BIG WINNER!

The DATSUN SPL-310 1500 CC Sports Convertible has made the world of competition motoring sit up and take notice. With 33 First-in-Class and 13 First Overall victories to its credit at the end of the '64 season, the SPL-310 has proved its stamina and superior handling on the track. In street form, it's fun and easy to drive, and delivers with more built-in extras than any car in its class. Standard equipment includes: racing steering wheel, tonneau cover, white sidewalls, transistor radio, heater/defroster, roll-up windows, full instrumentation, windshield washers, back-up lights, and a new, easily operated all-weather top.

Test drive the 1965 SPL-310 at your DATSUN Dealer's, and join the growing list of DATSUN enthusiasts.

Specifications — DATSUN SPL-310 1500 Sports Convertible

ENGINE	**Twin Carburetors**
Engine, no. cyl/type	4/O.H.V.
Bore/Stroke, in.	3.15/2.91
Displacement, cc.	1488
Equivalent cu. in.	90.7
Compression ratio	9.0:1
BHP @ rpm	85/5600
Torque @ rpm/lb. ft.	4,400/92
Carburetor, no./make	2/HITACHI
DRIVE TRAIN	
Clutch, dia. in. & type	8/single plate dry disc.
face area, sq. in.	19
Gear ratios	
4th	1.000
3rd	1.328
2nd	2.140
1st	3.515
Reverse	4.597
Synchromesh	2nd, 3rd, 4th
Differential, type & ratio	hypoid bevel, semi-floating/3.889

CHASSIS & SUSPENSION

Frame type—Pressed steel box section with x member

Brake type—Hydraulic. 2 leading shoes on front, leading and trailing shoes on rear; brake drum diameter 9 ins.

Tire size—5.60 x 13 4 p

Steering type & ratio—Cam and lever type; 2½ turns to lock; 3-spoke racing wheel, diam. 15.7 ins./14.8:1

Front suspension—Independent wishbones, coil spring with aircraft type double-acting hydraulic shock absorbers.

Rear suspension—semi-elliptic 4-leaf springs. Aircraft type double-acting hydraulic shock absorbers.

GENERAL	
Weight, lb.	2,006
Seating capacity	2 persons
Wheelbase, in.	89.8
Tread, front/rear	47.8/47.1
Fuel tank capacity, gal.	11.3
Overall length, in.	155.6
Width	58.9
Height	51.4
Ground clearance, in.	6.7

INSTRUMENTATION

Instruments—Speedometer (120 mi.) with mileage recorder; 6000 rpm tachometer (red-lined at 5400 rpm; ammeter; gas gauge; water temperature gauge; oil pressure gauge; ignition switch; starter switch; light switch; windshield wiper; choke control knob; map lights; electric clock; fresh air controls; fog lamp switch.

Warning lamps—Main beam warning light; amber turn signal pilot lamps.

ACCESSORIES

Included in list price—Electric 2-speed windshield wipers; windshield washer; cigarette lighter; dual horns; inside and outside rear-view mirrors; tonneau cover for the cockpit; adjustable deep-cushioned bucket seats; all-vinyl interiors; white sidewall tires; radio; heater/defroster; back-up lights; safety belts; roll-up windows; turn signals; locking glove box; center console with ash tray.

Optional extras—All accessories are included in deluxe package at no "extra" cost.

MAINTENANCE

Crankcase capacity, gal.—0.8.

Oil filter type—Gear type oil pump, full pressure feed, full flow oil filter, replaceable paper element.

Tire pressure f/r. lb.—22/22.

Cooling system, type/gal.—Pressurized radiator (4 lb. cap.) Centrifugal vane type water pump. Pellet type thermostat. 4 blade fan/1.7.

ELECTRICAL SYSTEM

Battery/amps—12v/4ah.

Generator wattage—300 watt alternator with voltage regulator.

Starting motor, hp.—1.4.

*12,000-Mile or 12-Month Warranty on all DATSUN cars & trucks.
(These specifications are subject to change without notice.)

NISSAN MOTOR CORPORATION in U.S.A.
Regional Offices

221 Frelinghuysen Avenue,
Newark, New Jersey 07114

1501 Clay Avenue,
Houston, Texas 77003

137 E. Alondra Boulevard,
Gardena, California 90247

Nissan touted the race track record of the "DATSUN SPL-310 1500 CC Sports Convertible" in a sales brochure that included other models, the "Four-Ten 4-Door Sedan" and "Four-Ten 4-Door Estate Wagon," the four-door models available with a "big sports package" that included a 4-speed manual transmission and bucket seats. *Automobile Quarterly Photo and Research Library* is a spring affair, and the MGB was not shown to the public until the London Auto Show in October 1962 (production of the MGB had begun in July, and the press got an advance look in September 1962). As the two cars were under development as the same time, similarities would seem to be coincidental, the effect of reading from the same page rather than one replicating the other's work.

The 1500's next appearance didn't come until the 1962 Tokyo show, late in the year and after the MGB's public showing. The car was slightly changed from New York, and it was in the form seen at the Tokyo show that it went into full production in 1963. Datsun immediately added the roadster to its U.S. lineup, which at the time included the 1200 cc Bluebird sedan, a pickup truck, and the Patrol sport-utility.

The roadster's appearance was, to use *Road & Track*'s later description, "not displeasing," though perhaps derivative of other marques. The editors noted bits and pieces that looked like similar bits on models of other marques—note the MG headlamp comparison—though one could do the same with just about any automobile.

Still, as the Datsun 1500 went into production, it was anything but groundbreaking. The steel body was mounted to a conventional chassis with box rails and an X-member. The front suspension was double A-arms with coil springs, tube shocks, and an anti-roll bar; the rear was a live axle on semi-elliptical springs with tube shocks. Tires were, per custom of the time, narrow 5.60-13-in bias ply, with cam and lever steering to direct where they were pointed. Brakes were drums all around. The 1500's "G-type" engine was an 80 bhp pushrod four.

The Datsun 1600 was immediately identifiable by the shorter side trim and its 14-inch wheels. *Automobile Quarterly Photo and Research Library*

That engine—certainly because of its heritage—would have looked at home under any British roadster's hood, except that the Datsun had an oversquare (3.15 x 2.91 in) bore and stroke. The cast-iron block and head, to the point, were fitted with a pair of 38-mm side-draft carburetors, actually Hitachi-licensed versions of the familiar British SU carb. The distributor, too, was Hitachi-made but a Lucas look-alike. The carburetors and exhaust of the non-crossflow head were on the left side of the engine, true to the car's right-hand-drive heritage. Datsun, however, installed an alternator, a recent innovation, instead of the traditional generator.

The transmission was a four-speed (synchro on the top three) with a floor-mounted shifter, with the clutch 8 in in diameter. Bucket seats were standard and, no doubt to placate the Japanese industry watchdog that frowned upon two-seaters, a third bucket was mounted crosswise behind the front passenger seat. It wasn't really suited to full-sized Americans, and no doubt Japanese passengers also considered it a penalty box. Not all 1500s had the seat, however.

Yet the interior was well done with good fit and finish. The dashboard, per traditional British custom, was a flat wood panel, with instruments flush mounted. The steering wheel had three metal spokes, with holes "drilled for lightness." A lower-case "d" graced the central horn button, the company not yet having adopted the circle and bar logo. The pedals were spaced to accommodate big American feet, though headroom was limited with the top raised. The top itself was purely in the tradition of British roadster, consisting of separate trunk-filling frame and fabric that only a Boy Scout could love. Once erected, however, the top was snug and watertight. So much for tradition.

The standard 1500 was well equipped, especially for a sports car. Heater (the passenger had separate controls), white sidewall tires, tonneau cover, back-up lights, seat belts, cigarette lighter, and transistor radio (still AM only) were all standard. *Road & Track* said, "we have never seen a car that comes with so many extras at no additional charge."

Road & Track also called the car "extremely easy to drive." That was written almost as a pejorative, confirmed by the magazine asking the opinion of several female staff members. The women reported it "easy to control and of a handy size." Male staffers were less enthusiastic, but fortunately the car wasn't tagged a "secretary's car," the ultimate epithet of the time.

Its 0-60-mph acceleration times were adequate, with *Road & Track* checking a 1964 model in at 15.5 sec, the quarter-mile coming up in 20.2 sec at 66 mph.

The 1500 was deemed to be still less than a true sports car, however, but rather a sport roadster. The magazine said the stiff suspension, while "not offensive at lower speeds, [when driven hard] begins to judder, the back end crabs around, the clutch loses its bite between fast shifts and the brakes are no more than marginally adequate for stopping from anything in the vicinity of the car's top speed." The Datsun was "favorably comparable" to similar models from Britain or the Continent that "make no pretense of being raceworthy sports cars in the accepted meaning of the term."

1600

The 1500 Sports was replaced in 1965 by the 1600 Sports. Though it's tempting to think otherwise, it consisted of more than just a larger engine, although even that wasn't done the easy way. Instead of a simple bore or stroke job, Datsun reduced the stroke (from 2.91 to 2.63 in.) and increased the bore (from 3.15 to 3.43 in). As a result, displacement went up to 1595 cc and horsepower was raised to 96 bhp. The bore and stroke changes allowed more flexibility, with peak power going from 5,600 rpm to 6,000 rpm. Quarter-mile times edged down to 19.9 sec and 0–60 mph to 13.3. The "R-type" engine was improved at engine number R-40001, changing from three main bearings to five.

Other mechanical changes included disc/drum brakes, with 11.2-in discs up front, replacing the all-drum setup of the 1500. The 1600's rear brakes were initially 9-in cast iron drums, replaced by like-sized Al-fin units on the 1967 model year. The gearbox, though still a four-speed, was a new unit and synchro on all forward gears. *Road & Track* described it "as good as any 4-speed we've ever driven," calling it "quiet, the linkage crisp and light, and the synchromesh faultless."

Changes to the exterior were subtle, but Datsun watchers could tell the difference without reading the displacement badges on the fenders or rear. Instead of the 1500's flat rectangular-pattern grille, a new deeper grille with a horizontal theme was used. The front fender flares were enlarged, and as a result, the chrome side strips extended forward only to the rear of the front fender flare instead of all the way to the front of the car. Badging was changed as well. On the 1500, the fender emblem "Fairlady," or sometimes "Datsun," appeared above the side spear just ahead of the door. On the 1600, the emblem "DATSUN 1600" was set above the front of the spear, changed in 1967 to "DATSUN" above the spear, with "1600" below.

The "vent windows" became functional

Bob Sharp won Datsun's first SCCA national championship, driving an F-Production 1600 roadster in 1967. *Automobile Quarterly Photo and Research Library*

on the SPL-311, actually opening (they had been fixed on the 1500), though for 1968, when the windshield was made taller, they no longer opened. Headrests were also added for the 1968 model year.

The hook-type trunk handle of the 1500 was replaced by a simple key-turn lock and the external trunk hinges of the 1500 were gone. The 14x5-in wheels replaced the 13s, with whitewall 5.60-14-in tires standard. Wheel width changed to 5 1/2-in for 1967.

A new convertible top, a foldback-type that didn't have to be stowed in the trunk, was covered by a tidy boot. The third seat was gone, replaced by a lightly padded shelf. The interior was also upgraded with a console, including a locking map box, over the driveshaft tunnel. The instrument panel was revised, too. Instead of a panel with four equal-sized dials with clustered minor instruments, the new arrangement had a large speedometer and tachometer mounted directly on the dashboard with smaller gauges also arrayed across the dash.

With no changes to the suspension, the ride remained "classic." The springs were still hard and the suspension travel limited, particularly at the rear. But on smooth roads, *Road & Track* called it "predictable to drive close to the limits," and "a near ideal car to learn to drive sports car style."

Racer Ken Miles tested six sports cars with *Car and Driver* in 1966 and found the suspension of the 1600 wanting: ". . . in some respects the Japanese have entered dark areas they don't yet fully understand. There is nothing in the automotive art so subtle as suspension design—and the Japanese don't have the hang of it yet. The Datsun is incredibly well-put-together, but the suspension is far too jolting, and to no advantage in road-holding. The ride is full of odd cycles and frequencies that are as cacophonous to our senses as Japanese music." It wasn't particularly fast compared to the competition, though its brakes were praised, and Miles turned quicker laps times around New York National Speedway's road course with everything but a Fiat 1500 sports car (which was soon to be replaced by the Fiat 124). The editors placed the Datsun fifth overall among the six, ahead of only the Triumph TR4A.

A comparison test by *Road & Track* in 1968 against a different set of sports cars had the Datsun faster in acceleration than everything but a Triumph Spitfire but, said the editors, "As we've said before, dollar for dollar, the Datsun 1600 offers more for the money any other sports car."

It had already been proven in competition as Connecticut car dealer Bob Sharp, whose name would become a fixture in SCCA and IMSA racing, won the SCCA F-Production national championship in 1967 in a Datsun 1500. Take that, Ken Miles.

Though the Datsun 2000 was introduced in 1968, Nissan continued to build the 1600 and sell it alongside its new, faster sibling until both were forced out of production by the 240Z. Said the trade publication *Automotive News*, "The two cars—Datsun 1600 and Datsun 2000—are direct market victims of the blazing success of Datsun's 240-Z sports coupe." All three were built in the same plant, and demand for the new car "led Nissan to decide to suspend production of the roadsters and turn over that plant's entire production capacity to the 'Z-Car.'" As a result, the two roadsters were absent from the 1971 model lineup.

What to Look For

The Datsun 1500 and 1600 deserve a special place in automotive history. It would take the 240Z to become an instant classic and the 2000 Sports to make the chassis it shared with the 1500/1600 a truly dominant force in American sports car racing, but it was the 1963 debutante 1500, for all her inexperience, that made the big change for Datsun. The first Fair Lady looks now like a parody of a sports car. The 240Z was undeniably a great vault into the realm of the modern sports car, but the 1500 looked like, and became, a real sports car. It and the 1600 that followed proved that a real sports car needn't come from Europe.

Unfortunately for collectors, however, when the 240Z was introduced, the 1500 and 1600 became nothing more than old cars, with the 1500 particularly unloved. As a result, many were worn down to the quick—and beyond. Exposure—to weather and/or penny-pinching repair—took its toll. Many were lost, simply not worth the effort to save. Only 3,148 of the 1500 SPL-310s were sold in the United

States, fewer than the 1600 SPL-311 (23,609), or the 2000 SRL311, at 15,718.

Many that survived the initial winnowing have been restored to some extent or another, although cars are still found in barns and junkyards. Either way, rust may be a problem. Look in all the usual places and bring a magnet to check for clumsy Bondo jobs. Typical rust sites include the leading edge of the front door pillar and the rear quarter panels, but convertible tops mean that the floor might be rusted, too. Patch panels aren't available yet, and the cost of good replacement front fenders and rear quarter panels can be real budget busters.

Check the operation of the top, particularly the front bow. Overly enthusiastic attempts to pull the top to the windshield header can bend the fragile bow. Of course, these are rare and replacements aren't readily available; finds will be expensive.

Surprisingly, upholstery and trim parts are available from several sources. Cost will be higher than for, say, parts for an MGB, as there just aren't as many Datsun roadsters to share the cost of development. Don't rule out dealers or general aftermarket vendors for assorted mechanical parts; over time, however, these sources will carry fewer of these parts as demand decreases.

1964 Datsun 1500 Sports (SPL-310)
Base price: $2,465

General

Layout	rwd
Curb weight (lb)	2,030
Wheelbase (in)	89.8
Track, front/rear (in)	47.8/47.1
Overall length (in)	155.6
Width (in)	58.9
Height (in)	50.2
Frontal area, (sq ft)	16.4
Ground clearance (in)	6.3
Seats	2+1
Trunk space (cu-ft)	2.6
Fuel tank capacity (gal)	11.4

Chassis and Suspension

Frame type	Box rails with X-member
Front suspension	Double A-arms, coil springs, tube shocks, anti-roll bar
Rear suspension	Live axle with parallel semi-elliptical springs, tube shocks
Steering type	Cam & lever
Steering ratio	14.8:1
Tire size (in)	5.60x13

Brakes

Type, front/rear	Drum

Engine

Type	OHV 8-valve Inline 4, cast-iron block and head
Bore & stroke (mm)	80.01x73.91
Displacement (cc)	1,488
Compression ratio	9.0:1
BHP @ rpm	85 @ 5,600
Torque @ rpm (ft-lb)	92 @ 4,400
Fuel supply	2 Hitachi 38 mm sidedraft carburetors
Fuel required	Premium

Drivetrain

Transmission speeds/types	4/manual
Gear ratios (:1)	
1st	3.515
2nd	2.410
3rd	1.328
4th	1.000
Final drive ratio	3.89

Performance (Source: Road & Track 1/64)

0-60 mph (sec)	15.5
1/4-mile, (sec @ mph)	20.2 @ 66
Top speed (mph)	87.5
Fuel economy (mpg)	26-31 (observed)

Chapter 4

★★ 1/2	**1968-1973 510 2-door**
★★	**1967-1972 510 4-door**

1967-1973 Datsun 510

When the first Datsun 510 rolled off the pier and onto a ship bound for America, certainly no one suspected that a cult was being initiated. There was little apparent to suggest it. The package was a conventional three-box design and the suspension and engine, though advanced for an economy sedan, didn't seem the stuff of legend. The independent rear suspension was a nice touch, particularly at the time for an economy sedan, but still not something to inspire a vogue with such longevity.

Yet thirty years later, the 510 is desired like few other sedans, economy or otherwise. Certainly the Toyota Corolla, for all its millions of

Introduced in 1968, the Datsun 510 was originally available only as a four-door model. *Automobile Quarterly Photo and Research Library*

A station wagon 510 debuted in 1969, but the new two-door caught the attention of racers and enthusiasts. *Automobile Quarterly Photo and Research Library*

sales around the world for so many years, doesn't inspire affection as does Datsun's surprising sedan.

Much, of course, has to do with competition of race-prepared models. The 510 was not a sports sedan as sold, nor even close. It did have fully independent suspension, with MacPherson struts at the front and semi-trailing arms at the rear, which gave it a very smooth ride compared to its contemporaries, most of which had live axles on leaf springs. The semi-trailing arms were attached to a subframe rubber-mounted (for noise reduction) to the unit body. And because there was no other suspension member that affected wheel location (coil springs and shocks mounted on the body), there was significant camber change with vertical wheel travel, similar to but not as severe as a swing axle. The station wagon didn't share the rear suspension used on the sedans, but rather had a live axle on leaf springs. Heavy loads would have played havoc with the semi-trailing arm-geometry.

A noise-reduction innovation on the sedans was the rear mount for the differential. The front was rigidly mounted to the suspension subframe, but the rear was bolted to a foot-long leaf spring that allowed longitudinal motion of the differential.

Standard tires were whitewall Toyo 5.60-13 bias ply, which must have been inexpensive but didn't help cornering, wet road handling, or hard braking. Most owners would sooner or later switch to radials, 175-13 being the recommended size. Tire size was limited by the standard wheels, 13x4-in steel discs, complete with gaudy wheel covers.

In stock condition the 1,595-cc L-16 engine wasn't particularly racy. In fact, it had a longer stroke than the 1.6 liter of the 410 it replaced, though the new engine had a chain-driven overhead cam instead of its predecessor's

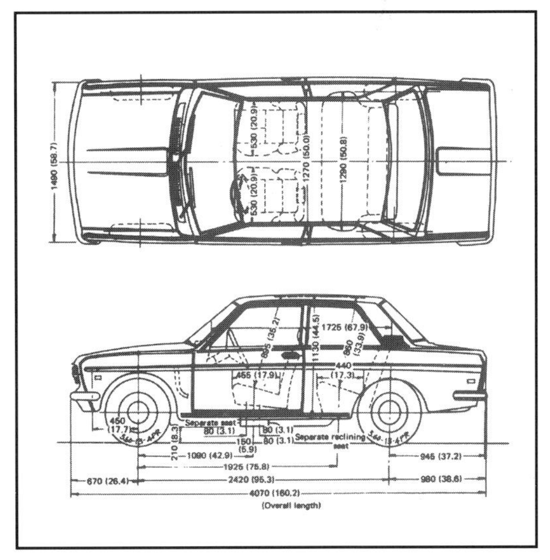

The dimensions of the two-door 510 were compact but efficiently used. *Nissan*

Connecticut Datsun dealer Bob Sharp added a 510 to the list of Datsuns he raced, here before the distinctive paint scheme later adopted. Note the 1970-type grille with the offset Datsun badge. *Automobile Quarterly Photo and Research Library*

pushrods. The engine actually had little in common with the earlier sedan's, and was about 20 lb lighter, sharing the concept—if not parts—with the U-series overhead-cam 2.0-liter engine of the 2000 roadster. (The OHC engines undoubtedly benefited from the annexation of Prince Motors, which had much more experience in overhead-camshaft engines than Nissan, which until that time had only built pushrod engines.)

It also didn't make near as much power as the roadster's 2.0-liter. Like the performance SSS version of the earlier 410, the 510 was rated at 96 bhp using a single Hitachi progressive two-throat carburetor. (A twin-carb SSS version of the 510's engine was available in Japan but was not officially imported to the United States, though some surely made it in. That engine was rated at 109 bhp at 6,000 rpm.) The engine had a cast-iron block and an aluminum head. The valves were all in line and operated via rocker arms passing under the camshaft. Intake and exhaust were both on the left side of the engine, accommodating the steering column of the home-market right-hand-drive cars but later making clearance problems for those who would install side-draft carburetors on U.S. left-hand-drive cars.

A four-speed all-synchromesh transmission was standard and a three-speed automatic was optional. At first, a Borg-Warner automatic made in England was used, though supply problems shifted sourcing to a similar unit made in Muncie, Indiana. Appropriate for a sedan, the shifter was column-mounted. The B-W Type 35 transmission performed well enough for most but it was rather primitive, requiring full throttle for downshifts, and with a rather tall first-gear ratio made the Datsun rather slow off the mark. The transmission, a $190 option, also absorbed much of the little engine's limited power, sapping acceleration and clipping fuel economy.

Datsun switched in April 1971 to a three-speed automatic made in Japan by JATCO, a joint venture of Nissan, Toyo Kogyo (Mazda), and Ford. The new automatic had its shifter mounted on the floor. Final drive differed depending on transmission and body style, with the manual gear box cars first coming in at 3.70:1 (later changed to 3.90:1), the sedan with automatics at 3.90, and the automatic wagons at 3.89.

The exterior styling was elegant and simple beyond the talents of most Japanese designers of the time; it bore a surprising similarity, in fact, to the 1963-66 Prince Skyline, and Prince was known to have employed Italian designers.

The interior, save for individual front seats, had classic "American" styling. A horizontal speedometer dominated the instrument panel and a hard plastic steering wheel had a padded cross piece with a semi-circular horn ring. The chair-height seats were vinyl covered and offered little lateral support. The optional air conditioning was an add-on under-the-dash unit. By present standards, the interior was dreadfully spare, with lots of painted metal, vinyl floor mats with a carpeted transmission hump, a cardboard package shelf, and a tall shift lever sprouting directly from the transmission hump. Console? There was no console. Reclining buckets were available in the four-door beginning with the 1971 model year, and in the two-door for 1973 only.

Interior room, however, was outstanding and adequate for trips by four adults. The trunk was surprisingly roomy as well, even if equipped only with a rubber mat for its floor and a cardboard cover for the fuel tank between the wheel wells.

Nissan advertised the four-door 510 just under two grand, at $1,996. At that it was quite a bargain and, along with the Toyota Corolla, supplied an answer to a question that hadn't quite been asked on the American market. But in 1982, *Road & Track's* Jonathan Thompson named it one of "the most significant cars of *Road & Track's* 35 years."

A two-door version of the 510 sedan wasn't added to the Datsun lineup until January 1969. Nissan advertised the two-door 510 as the "Datsun/2," a name that obviously didn't stick, with a list price of $1,935 that included, it was noted, "factory anti-freeze."

The grille was changed for the 1969 model year, and then again for 1970, the smaller horizontal bars painted matte black with a "Datsun" badge offset to near the left headlamps. Side markers changed for the 1969 model year, with a small amber wedge in front and a red button at the rear replaced by larger, more rectangular lamps incorporating a reflector.

The instrument panel was updated in 1970 to a much more modern-looking arrangement with deep-set gauges in a contour plastic panel. The left binnacle housed a round 120-mph speedometer, the smaller center hole held the fuel and temperature gauges, and the right had a round ring circled by ammeter, oil, and brake warning lamps. This latter, however, could be replaced by a dealer-installed 8,000-rpm tachometer that retained the warning lamps on its face ($60 then, very desirable now).

The 510 was available through 1973, though only in two-door form in the last year. For the duration, the engine wasn't significantly changed, rated at 96 bhp SAE gross—although in 1972 the switch to SAE net ratings changed the rating to 81 bhp with torque at 85 ft-lb.

What to Look For

It's hard to say where you'll find a Datsun 510. Some, of course, are still in normal day-today usage, others have been raced and rallied—in some cases for decades, and still others have been converted into street racers—in many cases faster than the famous Trans-Am cars of yore. Note that *Road & Track* tested the BRE Trans-Am 510, clocking a 15.2-sec/94.0-mph quarter-mile with road-race gearing and set-up. A factory SSS engine kit that included a dual-SU plus intake manifold was even covered by warranty if dealer-installed, though emissions regulations would seemingly have prohibited such modifications.

One thing is fairly certain, however. You'll find very few daily use cars in the Northeast and westward toward Chicago, or anywhere where de-icing salt is used on the roadway. The 510 was a notorious ruster. Those lightweight panels that worked so well for performance and economy were very vulnerable to corrosion. The rear wheel lip, the rocker panels (especially toward the rear), lower door edges, and near the base of the A-pillars are particularly susceptible. By now, however, almost all 510s sold new in the Rust Belt have returned to the smelter.

To find an unmodified 510, go to the Southwest. More were sold there to begin with, and a much, much higher percentage have survived. Still, after a quarter of a century, don't expect to find used-car lots full of them.

Mechanically, the 510 has earned a reputation for sturdiness. Although it's not impossible to break one, the 510's engine will easily top 100,000 miles; given normal care the engine and drivetrain will outlast the body. The aluminum head will warp easily if overheated, however. Unless the engine has been extremely modified, the transmission and differential are bulletproof.

Race and rally 510s are everywhere. Just check the classifieds in *Sportscar*, the official magazine of the SCCA, under the IT or GT

Bob Sharp drove to a second-consecutive national championship in B-Sedan in 1971. *Automobile Quarterly Photo and Research Library*

competition car listings. Better yet, go to SCCA regionals to see 510s with up to 25 years of competition on their logs—usually more than most of their drivers' logbooks, if not their birth certificates.

A number of 510s have survived as "cafe racers," modified for street performance with an availability of speed parts seemingly second only to the small-block Chevy. If fact, there's little interest in 100-point restorations of 510s. The excitement of the 510 is in what can be done with it. Buying an already modified 510 is a risky proposition, even more so than buying an unmodified car. Not only should the base automobile be subject to close inspection, but so should the quality of the parts used in modifying it *and* the quality of installation.

In standard stance, the 510 looks as if it's on its tippy-toes, and handling will be significantly improved if the car is lowered. This doesn't present a problem at the front, but shorter springs at the rear will put negative camber into the rear wheels. This doesn't hurt handling, but it will cause quick and uneven rear tire wear. The factory competition kit available in the early Sixties lowered and cambered a 510's rear suspension so much that wide bias-belted tires (with bias plys and a fiberglass belt) then available didn't put all of its rear tread on the pavement! Be prepared to have a competent shop decamber the rear wheels or plan to replace rear tires on a very regular basis.

1968 Datsun 510 four-door sedan
Base Price: $1,996

General

Layout	rwd
Curb weight (lb)	2,130
Wheelbase (in)	95.3
Track, front/rear (in)	50.4/50.4
Overall length (in)	162.2
Width (in)	61.4
Height (in)	55.1
Ground clearance (in)	7.5
Seats	5
Trunk space (cu-ft)	11.4
Fuel tank capacity (gal)	12.1

Chassis and Suspension

Frame type	Unit steel
Front suspension	MacPherson struts, lower lateral arms, drag struts, coil springs, anti-roll bar
Rear suspension	Semi-trailing arms, coil springs, tube shocks
Steering type	Recirculating ball
Steering ratio	15.5:1
Tire size	5.50-13
Wheel size (in)	13x4 diameter

Brakes

Front type	Disc
Size (in)	9.1
Rear type	Drum
Size (in)	9.0 diameter

Engine

Type	Inline 4
Bore & stroke (mm)	83.0 x 73.3
Displacement (cc)	1,595
Compression ratio	8.5:1
BHP @ rpm	96 @ 5,600
Torque @ rpm (ft-lb)	100 @ 3,600
Fuel supply	One Hitachi 2V

Drivetrain

Transmission speeds/types	4/man and 3/auto
Ratios 4-speed (:1)	
1st	3.38
2nd	2.01
3rd	1.31
4th	1.00
Final drive ratio	3.70

Performance (Source: Road & Track, 3/68)

0-60 mph (sec)	13.5
1/4-mile (sec @ mph)	19.7 @ 69
Top speed (mph)	98
Fuel economy (mpg)	23-27 observed

Chapter 5

★★★★ **1968-1970 2000 Sports (SRL-311)**

1968-1970 Datsun 2000 Sports

"Insignificantly heavier, moderately more expensive, but abundantly more powerful." That's how *Road & Track*'s road test summed up the new 2.0-liter overhead-cam Datsun 2000 Sports in its headline. The 2000 was essentially a bigger-engined version of the 1600. The body wasn't changed, and there were no changes to brakes or suspension. There was a new grille, smaller wheel covers, and minor trim changes, particularly a change to badges reading "2000."

The real news, of course, was under the hood. The overhead-valve engine had its pushrods and rocker arms replaced by a chain-driven overhead cam, and the displacement

The 2000 roadster had a new grille and had "2000" badging on the front fender (1970 model shown).

The rear end of the Datsun roadster was cut off flat. Back-up lights were factory add-ons under the rear bumper.

was increased from 1595 cc to 1982 cc by leaving the bore at 87.2 mm and stretching the stroke from 66.8 to 83 mm. The block was redesigned to reinforce the bottom end and a heftier crankshaft used to handle the higher loads. Nissan also enlarged the two carburetors from 38 mm to 44 mm to accommodate the increased displacement.

Nissan's first overhead-camshaft engine, the U-series, proclaimed "OHC" on its cam cover.

According to Nissan, the overhead cam arrangement wasn't used to increase maximum rpm but rather to improve the engine's durability at the top end. In fact, both 1600 and 2000 developed maximum power at 6,000 rpm. The 2000, however, made quite a bit more power. Instead of the U-16's output of 96 bhp, the U-20 was rated at 135 bhp, making 145 ft-lb of torque instead of 103 ft-lb. Do the math and it comes out to a *40 percent* increase in horsepower and torque!

Still not content, Nissan also added a five-speed gearbox to the powertrain. Common now, a five-speed gearbox was then almost revolutionary, as *Road & Track* had to explain to readers that fifth gear was "actually an 0.85:1 overdrive" and that its position in the shift pattern was "up and away to the right (as on the Alfa)—which seems the logical place for any 5th gear, Porsche and Ferrari practice not withstanding." Nissan changed all the ratios in the gearbox, making first through third taller (fourth was direct), and also made final drive taller as well, all possible thanks to the new engine's torque.

Even with the new gearing and a dozen or so more pounds, the 2000 could thoroughly skunk the 1600, doing the quarter-mile 2.6 sec faster. Zero-to-60 was a whole 3.1 sec quicker in the 2000. Top speed in fourth gear—fifth was a cruising gear—was 114 mph.

The chassis and brakes were up to the task of handling the 2.0-liter engine, at least if one were willing to endure the car's "classic" ox-cart ride. The tires were still Toyo 5.60-14 bias ply whitewalls and while these were slideable on a smooth surface; rough or rippled pavement would send the Datsun skittering toward the outside of the turn, especially at the rear. The brakes, unchanged from the 1600 and outstanding on that model, were no less so on the 2000.

A special competition engine kit, factory

Gauges in this 1970 2000 roadster included 140-mph speedometer, 8,000-rpm tachometer, and combination temp, oil pressure, amp, and fuel gauge in a cluster at right.

Introduced as 1968 models, 2000 roadsters should have had, by federal regulation, side marker lights. Some, like the one shown in this press release photo, did not. Perhaps these were made in 1967? *Nissan/Automobile Quarterly Photo and Research Library*

installed on a handful of roadsters in 1967, was available as a dealer option (or over the parts counter). Covered by warranty if dealer installed, the kit comprised a hotter camshaft and a pair of twin-throat Solex carburetors. So equipped, the U-20 engine was rated at 150 bhp and could push Datsun's roadster through the quarter mile in 16.8 sec and to a claimed top speed of 124 mph.

The SCCA classed the Datsun 2000 Sports in C-Production, up against Porsche 911s and such. It seemed rather stiff competition for the Japanese upstart, yet with encouragement from the company, the 2000 became a mainstay in the class in SCCA Regional and National competition, and remains so even until today.

The Datsun 2000 Sports was produced through 1970, but the runaway success of the 240Z doomed the 2.0-liter roadster (along with its 1.6-liter equivalent). With all three cars coming off the same production line and the 240Z in such great demand, the roadsters, which were getting long in the tooth anyway, were dropped.

What to Look For

As the Datsun 2000 Sports was mostly a 1600 Sports with a bigger engine, most of the advice in the preceding chapter applies here as well. One peculiarity of the U-20 engine was the tendency of its one-piece rocker arms to wear the camshaft lobes. The rockers rested on an adjustable post at one end and the valve stem at the other. The camshaft passed overhead, the lobe pressing the rocker arm, and thus the valve, downward. The rocker arm design was changed for the L-series engines used in the 510 sedans and Z-cars with the addition of a rubbing pad on the rocker's upper surface.

Datsun 2000 Sports (SRL-311)

Base price: $2,950

General

Layout	rwd
Curb weight (lb)	2,110
Wheelbase (in)	89.9
Track, front/rear (in)	50.2/47.2
Overall length (in)	155.7
Width (in)	58.9
Height (in)	51.6
Frontal area (sq-ft)	16.8
Ground clearance (in)	3.3
Seats	2
Trunk space (cu-ft)	n.a.
Fuel tank capacity (gal)	11.4

Chassis and Suspension

Frame type	Box rails with X-member
Front suspension	Double A-arms, coil springs, tube shocks, anti-roll bar
Rear suspension	Live axle with parallel semi-elliptical springs, tube shocks
Steering type	Cam & lever
Steering ratio	14.8:1
Tire size (in)	5.60x14

Brakes

Type, front/rear	Disc/drum

Engine

Type	SOHC 8-valve Inline-4, cast-iron block, aluminum head
Bore & stroke (mm)	87.2 x 83.0
Displacement (cc)	1,982
Compression ratio	9.5:1
BHP @ rpm	135 @ 6,000
Torque @ rpm (ft-lb)	145 @ 4,000
Fuel supply	2 Hitachi 44 mm sidedraft carburetors
Fuel required	Premium

Drivetrain

Transmission speeds/types	5/manual
Gear ratios (:1)	
1st	2.96
2nd	1.86
3rd	1.31
4th	1.00
5th	0.852
Final drive ratio	3.70:1

Performance (Source: Road & Track 11/67)

0-60 mph (sec)	10.2
1/4-mile (sec @ mph)	17.3 @ 79
Top speed (mph)	114
Fuel economy (mpg)	22-25 (observed)

Chapter 6

★★★★★ 1970-1973 240Z

1970-1973 Datsun 240Z

You almost had to be there. The impact the Datsun 240Z had on the market and on sports car enthusiasts was incredible. Everyone, it seems, wanted one, but there weren't that many to be had. It was an ideal seller's market. The list price was $3,526 but you couldn't get one for that price. Dealers added aftermarket wheels and tires, stripes, or just blatant mark-ups—or all three—to pad profits. And enthusiasts were willing to pay almost whatever was asked. At three-and-a-half thousand it was a bargain—even at 1970 prices. A Corvette 350 sold for about $5,000, a Volvo 1800E for around $4,500, a Jaguar E-type Coupe about $5,800. So if the 240Z cost $4,000 or even as much as $5,500, so what? Nobody else on the block had anything like it.

The 240Z was first seen in public at the Tokyo Motor Show in November 1969. There, for the Japanese market, it was still called the Fair Lady, with Z and ZL versions with a 2.0-liter inline six (from the home-market Datsun

The 1970 Datsun 240Z was stunningly good looking and is still handsome today.

Fastback styling of the 1970 Datsun 240Z was sleek and modern.

2000 sedan). Rated at 130 bhp, the difference between the Z and ZL was four- and five-speed transmissions, respectively. The top-of-the-line Z432 was powered by another 2.0-liter six, this one with four valves per cylinder and three twin-throat sidedraft carburetors, all good for 160 bhp.

These would not, however, come to America. Instead, Nissan sent the car with a 2.4-liter six derived from the four-cylinder engine used in the Datsun 510 sedan. Nissan took the modular approach and added two cylinders to the OHC 1.6-liter four. This "L-Series" engine, with cast-iron block and aluminum head, was new and fairly modern and fated to simplify the mishmash of engines developed by Nissan and those it inherited when it annexed car maker Prince Motors Ltd.

Nissan made the four into a six by stretching the cam and crankshafts by two cylinders. Many parts were identical. Using the same bore and stroke allowed the use of the same connecting rods and bearings. Pistons differed only in compression ratio, with the Z-car's flat top pistons having a higher ratio than the 510's pistons with the recessed faces. Some of the internal parts were shared with the home market's hot-rod 510 SSS, such as the larger exhaust valves; cam timing was also similar. It would prove to be a sturdy engine, with duplex-chain-driven camshaft and seven main bearings, safe to 7,000 rpm, with the tach's yellow band beginning at 6,500 rpm. Sustained high-rpm use, as in racing, would prove to be hard on crankshafts, however, as the long crank tended to whip, despite the seven bearings. Savvy teams considered crankshafts a consumable.

The engine leaned five degrees to the right, with exhaust and two carburetors, Hitachi-SU model HJG 46W (46 mm) on the left

Early 240Zs had cabin exhaust vents on the rear deck. Note the vertical defogger wires in rear window.

The 240Z's long 6-cylinder engine with twin Hitachi-SU carburetors had a look as classic as the exterior. The cam cover reads: "NISSAN 2400CC."

Early cars had 240Z badge on C-pillars.

side. The exhaust system, with muffler and resonator, was well-silenced, per Japanese taste.

Although a five-speed manual was offered on the home market and had been standard on the 2000 Sports roadster, the American 240Z buyer was limited to a four-speed box. A three-speed automatic and five-speed manual were promised, but the latter wouldn't appear on a U.S.-market Z until 1977.

All-independent suspension wasn't unheard of—Datsun's 510 had it—but it was still rare, especially in this price range. The front suspension was borrowed from the Datsun 1800 sedan, a home market car described as "a stretched and more luxurious car based on the 510." A subframe was rubber-mounted under the front of the unit body, to which the stamped-steel lower arm of a fairly conventional MacPherson strut system was attached. The strut had an internal tube shock absorber and coil spring. The lower arm was fixed fore and aft by a compliance strut, a rod rubber-mounted to the front of the body. A front anti-roll bar was standard.

The 510's independent rear suspension was by semi-trailing arms (*a la* BMW 2002) but the 240Z would get "Chapman struts." Named for Colin Chapman who used the system first on the Lotus Elan, Chapman struts are MacPherson struts moved to the rear. On the 240Z, long lower A-arms with a wide base provided the bottom link, while the strut with shock absorber and coil spring were mounted in a tower that intruded into the cargo area

Supplementary gauges in dash-top nacelles became a Z-car trademark. Note automatic-tuning AM radio and lack of air conditioning.

(and gave owners an opportunity to talk about independent suspension). The differential was rubber-mounted for noise control, with half-shafts with two constant-velocity joints and a ball-bearing spine per each to allow for wheel travel.

Standard tires were 175-14 radials. Standard wheels measured 14x4 1/2 in and were made of steel, though 5 1/2 in wide steel rims could be purchased from the dealer ($13.50 each). The wheels weren't "styled" but were fitted with full-disc mag-look wheel covers. Aftermarket "mags" were very popular, even when the dealer didn't use them to pad the price.

Vacuum-assist for the disc-drum braking system was used, with 10.7-in-diameter solid discs up front. The discs were actually smaller in diameter than the 2000 Sport's, but the single calipers had larger pads, providing the same amount of braking force. The rear drums (9.0 in inner diameter) were lifted directly from the 2000.

Despite all the good things underneath, it was the styling that meant love at first sight for most. Called reminiscent of Jaguar and Ferrari, the 240Z was instantly recognizable with a shape of its own. The "Z-car"—the nickname was immediate—combined smooth arcs with crisp edges. The bumper defined the lower radiator opening, the squared-off front edge of the hood the other opening, both coming to a slight but definite point. The headlamps were set into huge scoops (stylists would have preferred pop-up lamps) that became a Datsun Z trademark.

The hood swept back over a "power bulge" down the center, meeting the windshield at what seemed to be the mid-point of the wheelbase. The lower window line swept up to meet the fastback roofline, a "240Z" badge affixed to the C-pillar. There was just a hint of "Coke-bottle" in the rear fender. A large hatch with rectangular backlight (with vertical defogging wires) lifted for access to the luggage area. Large horizontally-split taillights were set in a recessed panel. For the Japanese, badging was remarkably restrained: a small "Datsun" disc on the nose, "Datsun" in bright letters on the front fenders just behind the wheel opening, the C-pillar badge, and a "Datsun 240Z" on the lower right corner of the hatch. At about 2,200 cars, the "Z" was changed from chrome to a white paint inlay (later to be a boon for those spotting early cars). The C-pillar badge was changed from 240Z to a stylized Z in a circle beginning February 1970.

Long doors aided entry, often a problem in low cars, though not that a true sports car driver would care. The interior fairly beckoned the enthusiast in. The dashboard was organically shaped, well ahead of its time, and featured two large, hooded nacelles, one for the 160-mph (dream on!) speedometer, the other for the 8,000-rpm tachometer, marked yellow from 6,500 to 7,000 rpm and red from there up. Centered on the dash but angled toward the driver were three smaller nacelles, the left two with tandem gauges (temp and oil, amp and fuel), the third with a clock. An "eyeball" vent was at each end of the dash.

In the center, the dash curved into an upper console with a vent, heater, and (standard) signal-seeking AM radio (with power antenna) controls. The shifter, with wood knob, chrome shaft, and soft leather-look boot, was in the center console, which had two levers on top and the "fly-off" hand brake lever mounted on the passenger side. The steering wheel had a "wood grain" plastic rim,

Except for the extravagant use of cheesy quilted vinyl, the early Z-car interior had a classic look, including a longish shift lever.

Before NHTSA chief Joan Claybrook dropped the maximum to 85 mph, the 240Z's speedometer read up to a very optimistic 160 mph.

hard and shiny, and three matte-black flat metal spokes with rectangular recesses apparently meant to evoke drilled holes.

The bucket seats were mildly contoured but covered in leather-grained vinyl, ventilated by perforated buttons. No one has ever seemed to like the peculiar quilted vinyl that covered the sides of the console and the shock towers in the luggage area. Wits compared the quilting to high-tech Chinese peasant garb. The luggage area was carpeted and had luggage straps but offered no security cover.

The reaction to the 240Z was phenomenal. The United States was expected to be the primary market for the sports car, and Nissan programmed 1,600 cars per month to be shipped stateside. By mid-1971 the quota was up to 2,500 per month in a market that could have supported 4,000 if Nissan had been able to build and ship them. Not only could Datsun dealers get more than list for a 240Z, but the *Kelly Blue Book* showed a retail value of a used 1970 240Z of $4,000! The 240Z wasn't sold in Germany until 1964 (when it was actually a 260Z), a time when 6,000 Zs were being built monthly—of which 5,000 went to the United States, 500 stayed in Japan and 500 went to the rest of the world. In the first year, the 240Z fell three short of 10,000 units sold.

1971

With sales going gangbusters, there was little reason for Nissan to make changes to the 240Z in 1971. During the year, modifications were made to the hatch to prevent water from leaking in: louvered vents on the hatch below

The hatchback was held up by a single gas strut and allowed access to more luggage space than was usually available in a sports car, even if everything was visible to passers-by. Note lack of C-pillar decoration in this early Nissan press release photo. *Nissan/Automobile Quarterly Photo and Research Library*

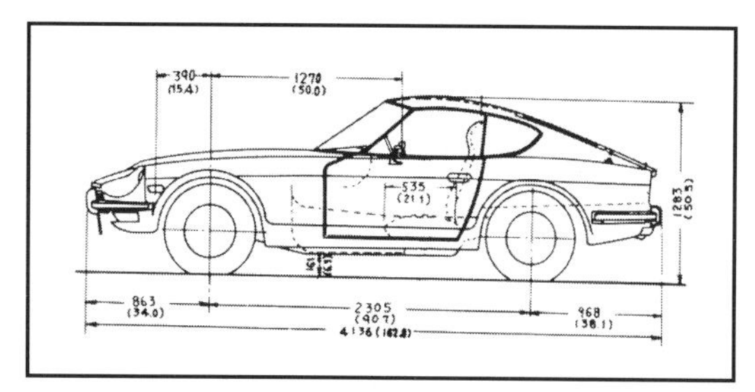

Drivers of the Z-car sat just in front of rear wheels. *Nissan*

the backlight were eliminated and cabin exhaust ventilation for the '71 Z was provided by a combination badge/vent (with the "Z" in white) replacing the '70's simple C-pillar badge.

Mechanical improvements included a new crankshaft for smoother engine operation. A new transmission case with improved bearings for the countershaft silenced complaints about noisy operation and also proved more durable.

The option of a three-speed automatic transmission was added at the beginning of the model year. The automatic, made by JATCO, a joint venture of Ford, Toyo Kogyo (Mazda) and Nissan in Japan, added $190 to the price of a Z. It had, at least, a floor shifter, a T-handle-topped unit replacing the manual lever on the center console. The transmission itself was a modern design, with "torque-demand" downshifts that didn't need full throttle. *Road & Track* found the upshifts "silky" but complained of whirring in its test car and occasional slippage. The testers also thought the throttle linkage was too "fast" for the automatic. Though no different than the manual's linkage, it made a smooth start harder with the automatic by opening the throttles too quickly.

The automatic transmission also upshifted too early under wide-open throttle acceleration. Left alone, shifts came at 5,000 to 5,100 rpm, well below redline, hindering performance. Shifting the automatic manually at 6,500 rpm improved acceleration, but still not up to the level of the four-speed manual.

The automatic also extracted a penalty in fuel mileage. While the four-speed manual got 21.0 mpg, the automatic got 2 mpg less. But overall, *Road & Track* considered the automatic and the 240Z a workable combination for those so inclined. Not too many were. Of 24,442 240Zs sold in 1971, 2,291 (less than 10 percent) were equipped with the self-shifter. The 510, in comparison, sold two-to-one manuals to automatics.

The 240Z had been built on the same assembly line as the Datsun 1600 and 2000 roadsters, but with production of the Z running six months behind orders, the older models lost their place on the line in 1971. That allowed the factory to gear up for more Zs, and with a full year's run, U.S. sales of the Z-car hit 26,733. The price had risen $70 (to $3,696), a modest increase considering the demand. Dealers still asked more.

1972-1973

The blue book (retail) price for a used 1971 240Z hit $4,400 by mid-1972, when the list price for a new 1972 240Z was $4,106, so forget dealer discounts! Nevertheless, sales continued to soar, up to 46,537 in 1972. At least for sales purposes, Nissan didn't need the five-speed manual transmission it had promised earlier. (But Zs exported to markets other than the United States received the five-speed as standard equipment beginning in 1972.)

The gear shift lever was relocated in 1972 (the transmission was changed from a F4W71A to a F4W71B beginning with September 1971 production), which also prompted the shifting of features on the console. Choke and seat belt warning lamps were added, though early 1972 models (through December 1971)

	Fully Automatic	Manually-shifted Automatic	Manual Transmission
0-60 (sec)	11.9	10.4	8.7
1/4-mile (sec)	18.2	17.6	17.1
@ mph	n.a.	82.0	84.5

had a blank where the belt lamp would go. The 1973 Zs, in response to new federal requirements, had new flame-retardant upholstery, but Z spotters are more likely to notice the federal bumper introduced that year. A rubber backing behind the bumper is another key if the slightly larger size isn't enough.

Pre-'72 240Zs had differentials "mounted too far forward," putting an angle of about 10 degrees in the universal joints in the axle half-shafts, resulting in early failures of the U-joints and a vibration that could be felt in the chassis in normal driving. Nissan responded by remounting the differential 1.37 in (35 mm) further aft in the 1972 models so that the axles came straight out from the differential. Parts for retrofitting earlier cars were also made available. To reduce roughness in the drivetrain, 1972 models had a one-piece driveshaft with a single universal joint.

Wheels were widened on 1972 models, up to 5-in-wide steel rims from the previous 4 1/2 in. Reclining seats were added in 1973.

Emissions regulations would be a problem for all car makers through the Seventies, and it was no different for Nissan. For 1972 the carburetor settings were leaned out, causing some drivability problems, but late in the model year an EGR (exhaust gas recirculation) system was added and the compression ratio was lowered to 8.8:1, which affected not only the way the Z drove but reduced performance and fuel economy as well. The engine's rated horsepower fell to 129 bhp at 6,000 rpm, and torque went to 127 ft-lb at 4,400 rpm. The drop in horsepower wasn't all "real," however, as 1973 was also the year that horsepower rating was changed from SAE gross to SAE net.

To meet emissions regs, these models also had a new Hitachi carburetor with a larger base that was closer to the exhaust manifold. The carbs absorbed more heat from the adjacent exhaust manifold, causing vapor lock and difficult hot starting. Aftermarket kits and homegrown solutions abounded, including adding insulation on the fuel lines, heat shields for the carburetors, vents in the hood, and even running with the front-hinged hood unlatched to let the heat out the back. Most effective, however, was the (illegal) installation of the earlier carbs.

The engine introduced in late '72 was carried over into the 1973 model year. Even with less power and more problems (although for the first time with reclining bucket seats), sales climbed, reaching 52,556 for '73. Nissan, though, was unwilling to let performance decrease, so the year would be the last for the 240Z. Sales for the three-plus years totaled 101,575. The '74s would address the loss of

Bob Sharp began his long affiliation with racing Z-cars with an SCCA National Championship in C-production. *Nissan/Automobile Quarterly Photo and Research Library*

power with the oldest solution in the books: More displacement.

What to Look For

Typical of the Seventies, the earlier cars are the best, though some prefer the 1971 models because of a tendency for the 1970 models to leak exhaust gases into the cabin via the taillights and rear hatch. The early 1972 models were little changed, but the effects of the lower compression ratio and other emissions equipment struck the 1973 models.

The inline six is tough. Reasonable care will result in remarkable reliability and longevity. Oil changes are vital, however, and laxity in lubrication duties will wear the cam lobes quickly. Poor performance from a Z that otherwise seems sound could be caused by a worn camshaft that doesn't fully open the valves.

Overheating is a common complication. Even when new, the 240Z's radiator was marginal. It's worse with a couple decades of scale accumulation. Cleaning helps, of course, but cars built 1974 and later have a three-core radiator which can be retrofit to solve the problem. The overheating also caused head gasket problems, the long aluminum head tending to warp when too hot.

The gear shifter should have a crisp, precise feel, but years of use will make it sloppy. This is easily remedied by a transmission overhaul. Take care shifting, however, as gearshift abuse may result in the gear lever coming adrift completely; it's held in place by a single retaining pin.

As this press release photo shows, the 1970 240Z came with wheel covers as standard equipment, though dealers often padded profits with mandatory aftermarket aluminum wheels. *Nissan*

Regarding rust, the 240Z has been called "second only to Fiat." Nissan used thin-gauge steel that will perforate quickly, and traps in the bodywork for moisture and dirt provide excellent sites for corrosion to start. A car that has lived its entire life in the rust belt is rare, though by 1973 buyers began rust-proofing Zs and keeping them inside when the salt trucks came out. Still, northern Z enthusiasts are advised to find that rust-free California car. Examine especially the rear fenders and rocker panels, though the floor and frame structure can also go bad. Check carefully: a rusty car is no bargain, whatever the price, though patch panels are available for common rust areas.

Problems with brakes include pads "frozen" to caliper housing by corrosion. Ironically, it's worst with cars that do a little winter driving: salt spray and dirt combine with inactive periods to leave pads or pad backing sticking to calipers. Regular use is a good antidote, as well as a thorough cleaning on a regular basis and fluid replacement every two years.

Clunking from the rear end may be loose universal joints. The 240Z has four more than cars with a live axle and they have a reputation for wearing faster. This is a relatively inexpensive repair, almost regular maintenance rather than restoration. Check the differential mounting bolts before replacing the U-joints, however. If they've loosened—which isn't uncommon—they'll cause noises similar to worn U-joints; tightening is cheaper than replacing.

A car with high mileage will surely have worn vinyl upholstery. Sun and exposure to the elements will also take their toll on the plastic. Upholstery kits are available from aftermarket vendors.

High-speed wander, if not excessive, is normal. Really, they all do that. Like the Stingray, the body was shaped primarily for visual appeal. That it had, but like the Stingray it also had substantial aerodynamic lift at the front and rear ends, making it a sort of poor wing. Racers countered this with front chin spoilers (such as Pete Brock's "Spook"—spoiler/scoop, which ought to be "spoop" but never mind) and rear deck spoilers. The racer look caught on with owners of street-driven Zs, who added these devices to their cars. They do work—at high speed—and if the right style can be considered "period correct" for anyone interested in a purist restoration.

As noted, another common period modification were "mag" wheels. These "custom" aluminum wheels were sold by the thousands to Z owners who wanted wider wheels—and anyway, wheelcovers were for Buicks, not sports cars. For the record, proper geometry requires a 1/2-in offset for 5 1/2-in wide wheels, 1/2-in offset for 6-in-wide wheels, and zero offset for 7-in-wide wheels.

Dealers also installed rear window slats for additional profit. It wasn't entirely unjustified, what with the large backlight contributing greatly to a mobile greenhouse effect at a time when air conditioning was not as commonplace nor as effective as today.

Model: 1970 Datsun 240Z
Base price: $3,526

General

Layout	RWD
Curb weight (lb)	2,238
Wheelbase (in)	90.7
Track, front/rear, (in)	53.3/53.0
Overall length (in)	162.8
Width (in)	64.1
Height (in)	50.6
Ground clearance (in)	5.7
Seats	2
Fuel tank capacity (gal)	15.9

Chassis and Suspension

Frame type	Unit steel
Front suspension	MacPherson struts, lower lateral arms, leading compliance struts, coil springs, tube shocks
Rear suspension springs, tube shocks	Chapman struts, lower A-arms, coil
Steering type	Rack & pinion
Steering ratio	17.8:1
Tire size	175-14
Wheel size (in)/type	14x4 1/2/steel disc

Brakes

Front type	Disc
Size (in)	10.7 diameter
Rear type	Drum
Size (in)	9.0x158

Engine

Type	SOHC Inline 6
Bore & stroke (mm)	83.0x73.3
Displacement (cc)	2,393
Compression ratio	9.0:1
BHP @ rpm	150 @ 6,000
Torque @ rpm (ft-lb)	148 @ 4,400
Fuel supply	Two Hitachi-SU 1v sidedraft carbs
Fuel required	87 octane

Drivetrain

Transmission speeds/types	4/man and 3/auto[1]	
Ratios (:1)	4-spd	Auto
1st	3.55	2.46
2nd	2.20	1.46
3rd	1.42	1.00
4th	1.00	—
Final drive ratio	3.36	3.54

Performance (Source: Road & Track 4/70)

0-60 mph (sec)	8.7
1/4-mile (sec @ mph)	17.1 @ 84.5
Top speed (mph)	n.a.
Fuel economy	21.0 observed

[1] Available 1971

Chapter 7

★★★★ 1974-1975 260Z, 260Z 2+2

1974-1975 Datsun 260Z and 260Z 2+2

For anyone who lived—and drove—through the decade, the Seventies were an era of diminishing expectations. Each new model year, beleaguered enthusiasts looked to see what had been done to their favorite cars in the name of "clean air" and faced the decision of whether to buy the newly emasculated version or grab one of last year's leftovers.

Car makers, of course, didn't like the situation any more than car buyers, and were even less eager to disappoint their customers. With emissions controls causing diminished power per cubic centimeter—as well as performance-robbing weight added for "safety" equipment such as 5-mph bumpers—manufacturers responded by making engines bigger in an attempt to hold the line.

Ergo the Datsun 260Z. As its name suggests, the engine was increased to 2.6 liters, or more exactly 2,565 cc, by stroking the L-series six. The bore stayed at 83.0 mm while the stroke was taken from 73.3 to 79.0 mm. It had the one intended effect, horsepower being restored to 139 bhp. But the power peak was lowered to 5,200 rpm, and although the markings on the tachometer weren't changed, the effective redline dropped to 6,000 rpm with the engine sounding, per *Road & Track*, "labored above 5,500."

The 1.6-liter L16 engine as used in the Datsun 510 had the same bore and stroke as the 240Z's 2.4-liter L24. However, Nissan increased both the bore and stroke to create the L18 used in the 1973 610. Meanwhile, the

The profile of the Z-car changed only slightly by making it a 2+2; note extra sheet metal between the door and the rear wheel arch. This 1974 260Z 2+2 has first edition of federal crash bumpers. *Automobile Quarterly Photo and Research Library*

The 1974 1/2 260Z 2+2 can be identified by its larger federal bumpers, larger grille, and repositioned front turn signals. *Automobile Quarterly Photo and Research Library*

260Z's displacement increase came only from a longer stroke, which, while increasing piston speed, still left the engine "oversquare," with more bore than stroke. The effect of the longer stroke still could be felt, however, in the reduced willingness to rev.

Reliability as well as emissions control were aided by a new distributor-contained transistorized breakerless ignition system. The two Hitachi-SU carburetors remained, however, still causing vapor lock and difficult hot starts. Yet performance was quick, but testers found the 260Z—while quicker than the later smog-equipped 240Zs—slower than the original Z. Lean surge, the scourge of '70s emissions controls, was still evident, and the 260Z suffered from dieseling (run-on), as much a sound of the decade as individual performances by John, Paul, George, and Ringo.

American enthusiasts looking overseas had even more reason to moan. Accustomed to European models losing horsepower to emissions controls on their trans-Atlantic trip, U.S. drivers could only be dismayed by the difference between the stateside and European-market 260Z. The Euro Z had 162 bhp, a five-speed manual, 3.7:1 final drive, 195R-70 tires, a front air dam and rear spoiler, and a curb weight of 2,425 lb.

For the States, the sports car's exterior was changed as little as necessary. Paint quality was improved, and all colors except red and white were metallics. The taillamp cluster was changed by making the back-up lights separate. Sports cars were given a one-year extension from full compliance with the 5 mph bumper rule that took effect in 1973, so the 260Z's barrier bashers, introduced on the last year's 240Z, weren't as bad as on many, and were fairly well-integrated into the design. The Z picked up 6 in in length, but mostly from the black rubber nubs that were visually rather unobtrusive.

The bumpers did add about 130 lb to the Z (one factor in the slower acceleration), so Nissan increased spring rates 20 lb/in front and 3 lb/in at the rear. As stiffening the front would normally increase understeer, a rear anti-roll

The short-lived 1975 260Z (front) and 260Z 2+2 were unchanged from 1974. *Nissan*

bar was added to counteract that effect and keep the 260Z's handling neutral. Shock absorber rates were not changed. Tire size wasn't changed either, so with more weight to handle, absolute cornering power decreased. Nissan attributed the car's reduced tendency to wander at speed to the weight of the front bumper combined with slightly longer rear springs giving the car more "wedge."

The instrument panel was tidied with bigger warning lights for choke and rear defogger, relocation of the stalkmounted lighting and wipe-wash functions, and integration of air conditioning into the heating/ventilation controls, mimicking GM's practice of the time. The contour of the dash was changed and the 240Z's steering wheel was replaced by one with a thick and grippy, padded vinyl-covered rim and a peculiar looking donut at its center. The door panels were reshaped, although the vinyl quilting on the driveshaft tunnel remained.

Washington imposed an odd safety device for 1974: the seatbelt interlock. It prevented a car from being started unless the seat belt was fastened, something that annoyed even the seat belt faithful, deterring the common habit of starting the car before buckling up or moving the car in the driveway without attaching the belt. But democracy had its way and the law was soon gone.

The yen-dollar relationship had its way with prices: the inexorable price climb put the base list price at a new high of $4,995. Nevertheless, *Motor Trend* called the 260Z, like its predecessor, "possibly the best sports car value in the world," echoed by *Road & Track*'s "one of the world's best sports-GT values."

260Z 2+2

Just as Jaguar broadened the market appeal of the Jaguar XK-E (or E-Type to purists) by adding a four-seat version of the coupe, Nissan supplemented the two-seat 260Z with a stretched version with a pair of occasional seats in back. The 260Z 2+2 had 11.9 in added to its wheelbase and overall length, and had its roof contour altered to fit a pair of contoured buckets where there had only been cargo space. The seats weren't very big—"occasional" as the "+2" indicates—with the driveshaft tunnel between them and the roofline limiting headroom to those about 5 ft 8 in or under *if* the front seat passengers moved their seats forward. The seats were really more appropriate for small children, making the 260Z 2+2 Dad's last sports car before the sedan.

No sheet metal changes were made forward of the cowl, and unlike the E-Type, the windshield was the same. The stretch was well done and not terribly noticeable from the outside. The rear side windows were reshaped, were larger, and flipped out for extra ventilation. Some 200 lb were gained in the transformation and spring rates were changed to compensate for the extra weight. Handling characteristics weren't significantly changed, however, and the ride was improved by the longer wheelbase.

Luggage room was increased, the rear seatback folding forward for a max of 9.2 cu-ft below the window line. With the seatback raised however, the two-seater wins, with 8.5 to 5.9 cu-ft.

That most buyers went for the two-seaters shouldn't be a surprise, considering the personality of the car. But the popularity of the 2+2 indicates that Nissan knew how to make the best of a good thing.

1974 1/2–1975

Late in the 1974 model year bumper size was increased to meet in advance the more severe 1975 federal pendulum bumper test. The grille was modestly enlarged for better cooling and the turn signals were repositioned above the big new bumper for better visibility. Technically still 1974 models, Datsun U.S.A. called these cars 1974 1/2 models. Otherwise unchanged from the 1974 260Z, they were another 6.3 in longer in overall length and had another 130 lb to haul about. The 1975 models were unchanged, and lasted only for part of the year, replaced mid-year by the new 280Z.

What to Look For

The 1974 260Z is the better car of the two years, at least by anyone's standards but an insurance agent's, as it has the smaller bumpers and therefore weighs less and looks better. Otherwise there's no difference, so you might want to put condition ahead of esthetics. Your call.

Mechanically and structurally, the 260Z was very similar to the 240Z that preceded it and so naturally much of the same advice applies. The 2.6 engine doesn't rev as freely as the 2.4, but that trades off against the electronic ignition and the displacement increase. It is possible, by the way, to make a 240 a 260 (or vice versa) merely by swapping the connecting rods and crankshaft, though there's really little motive to do so.

Rust is still a problem, though the 260Z isn't as rust-prone as earlier cars, thanks to running improvements by the factory. Examination of individual cars, however, will require no less vigilance, as exposure and care will have had a greater influence on their condition than the changes.

Despite its similarity to the 240Z, the 260Z isn't nearly as desirable to the collector. Doubtless there's the psychological effect of model name, similar to that which affects an "early" or "late" example of a given model. But while the 260Z wasn't as fast as the original 240Z, it was about par with the early 1973 Z-cars and quicker than the late 1973 240Z. Fit and finish were the best yet. And it sold well, more than 58,000 units in its two-year history. Nevertheless, from an "investment" standpoint, the "260" moniker hasn't worn—and won't wear—as well as "240." It may not be fair, but that's the way it is. But from a purchase standpoint, these cars should be cheaper to buy.

1974 Datsun 260Z
Base price: $4,995
Major options: Air conditioning, $400

General

Layout	rwd
Curb weight (lb)	2,665
Wheelbase (in)	90.7
Track, front/rear (in)	53.3/53.0
Overall length (in)	169.1
Width (in)	64.1
Height (in)	50.6
Ground clearance (in)	5.7
Seats	2
Trunk space (cu-ft)	8.5
Fuel tank capacity (gal)	15.8

Chassis and Suspension

Frame type	Unit steel
Front suspension	MacPherson struts, lateral arms, coil springs, anti-roll bar
Rear suspension	Chapman struts, A-arms, coils springs, anti-roll bar
Steering type	Rack & pinion
Steering ratio	18.0:1
Tire size	175HR-14
Wheel size (in)/type	14x5/steel disc

Brakes

Front type	Disc (vacuum assist)
Size (in)	10.7 diameter
Rear type	Drum
Size (in)	9.0 diameter x 1.6 wide

Engine

Type	OHC Inline 6
Bore & stroke (mm)	83.0 x 79.0
Displacement (cc)	2,565
Compression ratio	8.8:1
BHP @ rpm	139 @ 5,200
Torque @ rpm (ft-lb)	137 @ 4,400
Fuel supply	Two Hitachi-SU 1V
Fuel required	Regular

Drivetrain

Transmission speeds/types	4/man and 3/auto
Gear Ratios (:1)	
1st	3.59
2nd	2.25
3rd	1.42
4th	1.00
Final drive ratio	3.36:1

Performance (Source: Road & Track, 2/74)

0-60 mph (sec)	10.0
1/4-mi (sec @ mph)	17.9 @ 78.5
Top speed (mph)	113
Fuel economy (mpg)	20.0 observed

12+2 available

★★★1/2	1975-1978 280Z and 280Z 2+2

1975-1978 Datsun 280Z and 280Z 2+2

The Seventies must have been a trying time to make a sports car. Emission controls, however noble the cause, took resources that could otherwise make cars more fun and exciting. And bumper regulations—what would Thomas Jefferson have said? Regardless, rules were rules, and if you wanted to sell a car in the United States, you had to jump through all the hoops.

But selling cars is more than meeting regs. You also have to make a product sufficiently enticing for customers to want to trade a goodly amount of money for it. The 240Z, a knockout in 1970, would have been an also-ran in 1975 if encumbered with the "safety" bumpers and its 2.4-liter engine strangled by emissions controls. Nissan had responded with the 260Z in 1974 and in mid-1975 fired another shot with the 280Z.

Park a 1975 260Z next to a 1975 280Z and the only way to tell the difference—without opening the hood to see the fuel injection system—is by the model designation on the front fenders. *Nissan*

Again the name change represented an increase in displacement. Instead of increased stroke, the bore of the inline six was increased by 3 mm for a new total displacement of 2,754 cc. Despite an almost-Third World compression ratio of 8.3:1, output climbed from the 260Z's 139 bhp to 149 bhp.

The increased power, however, was from more than just increased capacity. Nissan abandoned the Hitachi-SU carburetors in favor of L-Jetrontic fuel injection, made by Nissan and Diesel Kiki in Japan under Bosch patents. The injection system used a "flap" to monitor air flow, and varied injector pulse length (all six injectors working simultaneously) as calculated by an electronic computer.

The fuel injection not only improved performance but drivability was better too, with lean surging, hesitation, and other evils all but exorcised. The engine seemed unresponsive at lower rpm and throttle openings, and a fuel cut-off under certain closed-throttle conditions exacerbated backlash inherent in the Datsun's drivetrain.

California cars got a catalytic converter for the first time, requiring unleaded fuel. These "CAT" cars can be identified by warning lights on the dash: "catalyst" to warn of converter overheating, and "floor temp" to warn of an overheated floor. (The driver's sun visor also carried a warning about parking over grass, weeds, or brush perhaps starting a fire—a real possibility.)

In copy on one of its PR photos sheets, even Datsun U.S.A. admitted that, externally unchanged, the only visible identifier for the 1976 model was the voltmeter replacing the ammeter on the center console. *Automobile Quarterly Photo and Research Library*

Gear ratios in the four-speed manual transmission were made taller, with first gear, for example, going from 3.59:1 to 3.32:1, to counteract a numerically higher final drive ratio, 3.36:1 changing to 3.55:1. There was still a fair amount of clutch-slipping to get off the line and the change in final drive meant higher-rpm cruising.

The net result was that the 280Z nearly matched the performance of the 1970 240Z—something most car makers could not say about their mid-Seventies' performance cars. It was a remarkable accomplishment, especially considering that the 280Z also used regular fuel (only unleaded regular was available by then) rather than the 240Z's requirement for premium.

The bigger bumpers from the 1975 260Z went on the 280Z (for which they were ultimately intended). These were grotesque, front and rear, and interfered with the original and unchanged body panels.

Bigger tires were finally fitted, 195/70x14 instead of 174x14, and although these were needed for the heavier Z-cars, they meant more work for the driver in parking and other low-speed maneuvers in the non-power steering cars. The weight increase also inspired the installation of larger struts.

The changes to the interior for the 260Z were carried over for the 280, although the quilted vinyl on the transmission hump and rear shock towers was finally gone, replaced by carpeting.

Cost, of course, had gone up once again, the two-seater now up to $6,284, the 2+2 hitting $7,084. It had been only five years since the introduction of the original 240Z and the price had almost doubled. Performance and fuel economy had both slipped slightly, and the car was beset with ridiculous bumpers. Yet these were the result of changing times, and Nissan had done well to maintain the car's original talents and spirit.

1976

The 280Z, introduced late in the 1975 model year, continued unchanged for the 1976 model year. Base prices, however, went to $6,594 for the two-seater, and $7,394 for the 2+2.

1977

Finally in 1977 there was a five-speed for U.S. Z-cars. It was an option—the four-speed was still standard and the three-speed automatic remained optional—but for only $165 more, it's hard to imagine why anyone would opt for the four-cog manual box.

Fifth gear in the new gearbox was an overdrive with a ratio of 0.86:1 and was best for making highway driving quieter and more fuel efficient. All other ratios remained the same as

Datsun boasted 18 engineering and styling changes on the 1977 280Z. Most noticeable were new hubcaps, functional hood louvers, and restyled bumpers. *Automobile Quarterly Photo and Research Library*

the four-speed's, and the final drive ratio wasn't changed, so there was no change in performance. But at 60 mph engine speed in gears dropped from 2,960 rpm in fourth gear to 2,545 rpm in fifth. *Road & Track*, in a test loop that didn't emphasize top gear, recorded a 2 mpg increase in fuel economy, from 19.5 to 21.5 mpg.

The increase in fuel economy was accompanied by an increase in horsepower. Major specifications, including displacement, fuel injection, and compression ratio, were unchanged, but Nissan was able to find enough stray ponies for a new output rating of 170 bhp at 5,600 rpm as well as 177 ft-lb of torque at 4,400 rpm. The jerk-producing fuel cutoff during coasting was still bemoaned by testers, however.

Profile of the 1978 280Z shows off the considerable size of the federal bumpers. *Automobile Quarterly Photo and Research Library*

The 1977 280Z was also 1/5 in longer thanks to a bumper restyle, though the functional hood louvers and new wheel covers were more apparent to most observers. Sound systems were also upgraded, *Road & Track* raving about its test car's AM/FM/8-track stereo unit. Also among 18 engineering and styling changes boasted by Datsun were breathable vinyl seats, a visor vanity mirror and an ash tray lamp.

1978

This was the final year of the original Z, and the last for the 280Z. With a new model in the works, changes were limited to items such as a new color (Sky Blue Metallic) and a standard AM/FM radio.

A special edition, a familiar maneuver by manufacturers to hype an aging model, featured black paint for the first time on a Z car. Called the "Black Pearl" 280Z, it had a black pearl metallic finish with red and silver striping on the sides, hood, rear deck, and around the headlamps. Offered only on the two-seater coupe, the Black Pearl package also included dual "racing" mirrors and sunshade on the backlight.

What to Look For

The last version of the first Z may be one of the best restoration prospects. Nissan had improved anti-rust design, and original buyers, alerted to experiences with earlier Z cars, were more likely to have had their Datsuns rustproofed. Add to that the fact that the tin worm has had fewer years to work on these later cars; more are likely to have survived.

Few mechanical differences from earlier models means the 280Zs are stout and reliable under the sheet metal. But it's not a 240Z, a name that has retained and gained magic over the years. 280Z is still a first generation Z car, though, and prices are still extremely reasonable, even for well-maintained or restored cars. And more than the 240Z, the 280Z is more likely to show up in local classifieds than in *Hemmings*—something that will affect prices. Effects of emissions and bumper regs aside, the 280Z is a refined version of the original Z. The added luxury is something that simply must be endured.

The 280Z, however, will never have the collectibility of the 240Z, the original, and prices reflect that now and will always.

1975 Datsun 280Z
Base price: $6,284
Major options: air conditioning, $425

General

Layout	rwd
Curb weight (lbs)	2,875
Wheelbase (in)	90.7
Track, front/rear (in)	53.3/53.0
Overall length (in)	173.2
Width (in)	64.2
Height (in)	51.0
Ground clearance (in)	n.a.
Seats	2
Trunk space (cu-ft)	n.a.
Fuel tank capacity (gal)	17.2

Chassis and Suspension

Frame type	Unit steel
Front suspension	MacPherson struts, lateral arms, coil springs, anti-roll bar
Rear suspension	Chapman struts, A-arms, coil springs, anti-roll bar
Steering type	Rack & pinion
Steering ratio	n.a.
Tire size (in)	197/70x14
Wheel size (in)/type	14x5/steel disc

Brakes

Front type	Disc (vacuum assist)
Size (in)	10.7 diameter
Rear type	Drum (vacuum assist)
Size (in)	9.0x1.6

Engine

Type	SOHC Inline 6
Bore & stroke (mm)	86.1 x 79.0
Displacement (cc)	2,754
Compression ratio	8.3:1
BHP @ rpm	149 @ 5,600
Torque @ rpm (ft-lb)	163 @ 4,400
Fuel supply	Nissan Bosch-type L-Jetronic fuel injection
Fuel required	Regular

Drivetrain

Transmission speeds/types	4/man and 3/auto
Gear ratios (:1)	
1st	3.32
2nd	2.08
3rd	1.31
4th	1.00
5th	
Final drive ratio	3.55

Performance (Source: Road & Track 6/75)

0-60 mph (sec)	9.4
1/4-mile (sec @ mph)	17.3
Top speed (mph)	119
Fuel economy (mpg)	19.5 observed

12+2 optional

Chapter 9

1979-1983 Datsun 280ZX and 280ZX 2+2

★★★	**1979 280ZX**
★★★1/2	**1979 280ZXR**
★★★1/2	**1980 280ZX 10th Anniversary**
★★★1/2	**1981-1983 280ZX Turbo**

Replacing a dud is easy. Just do something—*anything*. Replacing a successful model is something else. It can't be too much like the old model, or else why change? And if it's changed too much, the essence that made the old model a success may be ruined. On the other hand, times may have changed and the old formula may no longer be valid. What to do, what to do?

Nissan executives must have had fingers and toes crossed when they introduced the 1979 280ZX to the press at Portland International Raceway. The cars they brought—two-seater and 2+2 versions—resembled the older

The 1979 280ZX's massive restyle concealed a completely new chassis underneath and represented a change in philosophy at Nissan. Standard equipment wheelcovers are shown here. *Automobile Quarterly Photo and Research Library*

The 1979 280ZX could be ordered with optional alloy wheels, as seen here. *Automobile Quarterly Photo and Research Library*

models. The general profile was there, the scooped headlamps and hood bulge were carried over. There was a whole new car underneath, however. Basically, Nissan built the new ZX on the chassis of the 810 sedan—so where before they had advertised the six-cylinder 810 as having the engine of a sports car, they could have (but of course wouldn't) advertise the 280ZX as having the chassis of a luxury car.

The suspension was still fully independent and had MacPherson struts at the front, albeit with tension rods (locating rods ahead of the

The 2+2 version was continued on the 1979 280ZX, as shown here in GL trim. *Automobile Quarterly Photo and Research Library*

The 280ZX had fully independent suspension, though the rear used semi-trailing arms. *Automobile Quarterly Photo and Research Library*

The 1980 10th Anniversary 280ZX celebrated 10 years of Z-cars. It carried a special anniversary logo just behind the front wheel cut-outs. *Automobile Quarterly Photo and Research Library*

Turbocharging debuted with the 1981-1/2 280ZX Turbo, but only with an automatic transmission. *Nissan*

lower suspension arm) rather than compression rods (the rods behind the lower suspension arms). At the rear, however, the struts were changed to semi-trailing arms. No doubt the cost sharing made economic sense, but enthusiasts bemoaned the greater camber and toe changes that would affect handling.

The engine and drivetrain were largely carried over, though with various engineering details improved. The transistorized ignition was miniaturized and mounted on the distributor, the exhaust manifold was lighter and stronger, and the viscous-drive radiator fan was quieter and more efficient. The optional air conditioner included an auxiliary blower that pumped cool air onto the fuel-injection metering block and fuel lines to prevent vapor lock when the ignition switch was shut off and the coolant temperature was above 230 degrees Fahrenheit (F). Horsepower was down to 135 bhp, with maximum torque of 144 ft-lb at 4,400 rpm. California cars came catalyst-equipped and lost 3 hp. Nissan claimed only recalibration of ignition and fuel-injection systems.

The five-speed manual, formerly optional, became standard and ratios were slightly changed, final drive going from 3.54:1 to 3.70:1.

The 1981-1/2 Turbo Z had special badging and alloy wheels. *Nissan*

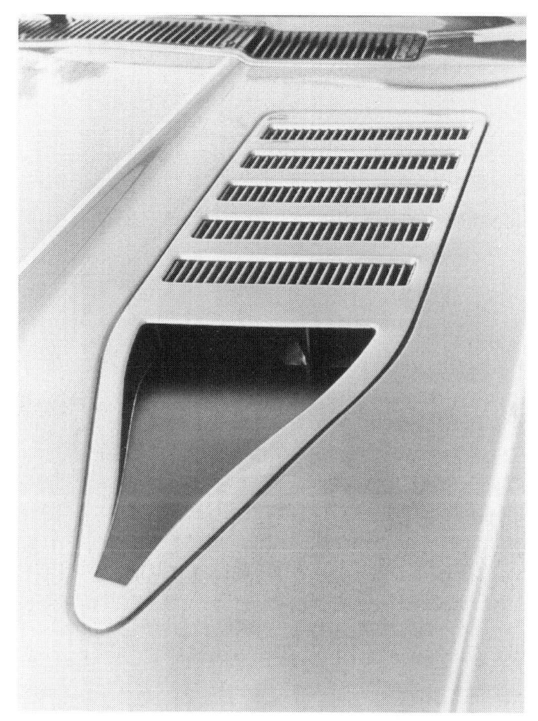

The functional NACA hood scoop was introduced on the 1981-1/2 280ZX Turbo. *Nissan*

The three-speed automatic continued as an option. The 280ZX got four-wheel disc brakes, ventilated at the front.

More apparent than the mechanical changes was the new skin. It was obvious that Nissan's stylists were borrowing heavily from the original design, but the car took on a whole new personality. The dimensions were nearly identical, with the wheelbase of the two-seater increasing by 0.7 in in length. (The ZX 2+2's wheelbase was 7.9 in longer than the two-seater's, making it shorter than the 2+2 Z's.) Height and length were about the same as the predecessor's while a couple of inches were gained in width, but the car looked bigger. The bumpers were more integrated into the new design, and so the sheet metal was actually longer than that of the original car. The sharper creases of the bodywork—a styling trend of the time—also probably made the car look larger as well. Surprisingly, it wasn't any heavier.

The 1982 280ZX Turbo had new alloy wheels and other styling changes. *Nissan*

The interior was markedly roomier, with more shoulder room and bigger seats. The back seats in the 2+2 were wider as well, though head and leg room were still more suited to children. Cargo room increased, too. But more noticeable was the upgrade of interior appointments. The inside was fully color coordinated, including the dash and instrument panel, and cut pile carpet was everywhere. The instrument panel was restyled with a large pod in front of the driver, with three nacelles for minor gauges to the right recalling the original design. The gauges got trendy red-orange lighting. Nissan provided illumination for everything, and there was even a lighted vanity mirror. The outside mirror was power adjustable, and power steering and air conditioning were standard on the 2+2, optional on the two-seater. A remote hatch release was standard on both.

The Grand Luxury package—can you imagine this on the 1970 240Z?—added alloy wheels (steel rims with cheesy wheelcovers were standard) and raised-white-letter Bridgestone tires; rear window wiper and washer; cloth upholstery (vinyl was still standard); door map pockets; power steering on the two-seater; power windows and passenger door mirror; door map pockets; AM/FM stereo radio with four speakers and joy stick ("surround sound") control; a central warning system; and a dual fuel gauge. The latter had two needles, one for the entire fuel tank and the

Urethane covering on the bumpers distinguished the 1982 ZX from earlier models. The model shown here is a 1982 280ZX Turbo 2+2. *Nissan*

other that read for the final quarter tank. *And* there was a low-fuel warning light. The central warning system, between the speedometer and tachometer, alerted the driver to burned-out brake lights, taillamps, or headlamps, and low fluid levels for the windshield washer, battery, or coolant.

The steering wheel was not sporty at all, being one of those devices with A-shaped spokes and no place to hang your thumbs in the 10-2 position. The shifter got a hard rubber boot, a sort of pleated pyramid, all very Seventies Modern.

Nissan put a lot of work into sound deadening, and the 280ZX was significantly quieter than its predecessors. Acceleration, despite the lower horsepower numbers, was about the same as the 280Z's, but critics found the handling was not what it had been. This was attributed in part to the new rear suspension's camber steering effect and also to softer bushings used for a more luxurious ride. Softer bushings holding the steering rack made steering less precise; the power steering was also criticized as "numb."

In effect, for the 280ZX, Nissan changed the emphasis in "sports-GT" from "sports" to "grand touring." That's not bad, merely different, and the question, from a business standpoint, was whether it would help Nissan's bottom line. It wasn't Nissan's job to make sports car enthusiasts happy, but to make money.

Nissan gave the 280ZX a base price of $9,899, a number that would go to five figures after a brief brush with the option list. The GL package was priced at $2,284, California emissions $105, automatic transmission $295, and two-tone metallic paint $99. For $399, a special edition called the ZXR was offered. Built in a lot of 1,000 to homologate its whale-tail rear wing for racing, the ZXR was identified by special badging (emphasis on the "R") and a

The 1983 leather/digital option featured a hard-to-read linear tachometer. *Nissan*

broad two-tone stripe down its side over the required Silver Mist paint.

1980

The Datsun 280ZX was all-new for 1979, but that didn't stop Nissan from bringing new features to the model in 1980. Most notable was a T-top—called "skyroof" by Nissan—which was necessary for the ZX to compete in the market with the Corvette, whose price the Datsun was only a ring or two on the cash register behind. Unlike the 'Vette's roof, which could be dislodged with a well-aimed thump of the hand, the Datsun had safety catches to counter the thief. The Datsun also had air deflectors that hinged out of the sun visors, though the T-top created a fairly draft-free cockpit without them. It proved to be popular: By the end of the year half of all 280ZXs had the option.

Other new options were leather upholstery ($300) and automatic temperature control for the HVAC. The thermostat cost $70 atop $635 for air conditioning. A Grand Luxury package cost $1,704 for 1980. The base price at the beginning of the year again was $9,899, but a July increase, blamed on "inflationary pressures in Japan and a significant change in the yen-dollar relationship," bumped the base price to $10,349.

While economics made a hit on price, emission regulations clipped another three horses from the engine's output, and the 1980 280ZX rated at 132 bhp at 5,200 rpm.

A new model introduced in 1980 was the 280ZX Anniversary Edition. This model made it's debut at the Chicago Auto Show in February and was finished in black with gold tape stripes trim, and gold-tone alloy wheels. It featured tinted glass panels for its T-top roof, headlamp washers, automatic climate control, and Goodyear Wingfoot radials. And of course a commemorative decal for the front fenders. Only 3,000 were built.

1980 280-ZX base prices can be split into early year and late-year listings:

	Early	**Late**
Two-seater	$9,899	$10,349
Two-seater GL	$12,238	$12,688
2+2 GL	$13,153	$13,603
10th Anniversary	$13,850	$14,300

1981

Prices edged up again in the fall of 1980 when the 1981 models were introduced, but at least horsepower was up as well. "Mechanical refinements," a three-way catalytic converter, and a compression ratio hiked to 8.8:1 raised output to 145 hp. Prices went to $10,699 for the two-seater, $12,503 for the GL, and $14,053 for the 2+2 GL (the 2+2 was available only in GL trim). The option of the T-top was extended to 2+2 models.

Nissan saved the real excitement for the April 1, 1981, introduction of the "1981 1/2" 280ZX Turbo. Turbocharging had become a popular hop-up technique for the aftermarket for several years, and in the previous two or three years manufacturers from Saab to Buick had begun to turbocharge models for performance and economy. The 280ZX was the first Nissan to use the exhaust-pressure supercharging device and it netted the inline six an increase to 180 bhp at 5,600 rpm. Torque was increased to 203 ft-lb at 2,800 rpm. An AiResearch turbo delivered the pressure, and though displacement was not changed, the engine conversion was extensive. The cylinder head and the shapes of the piston and rings were changed, new intake and exhaust manifolds were required, and the camshaft was revised. Compression was reduced from 8.6:1 to

Solid color (non-two-tone) 1983 Turbo-ZXs received a "blackout package" that, among other things, included "turbo" striping. *Nissan*

7.4:1, and the turbocharger was limited to 7.0 psi boost by a wastegate with a back-up pop-off valve. A larger radiator was used and an oil cooler was added, and the diameter of the catalytic converter's inlet and outlet pipes were increased to reduce back pressure.

The Turbo benefited from the growing use of engine control computers, getting one of its own with a typically Japanese acronym, ECCS, for Electronic Concentrated Engine Control System, which controlled fuel injection, ignition timing, idle speed, exhaust-gas recirculation, and fuel pump operation. The engine also had a knock sensor, with the engine retarding spark if knock occurred.

Apparently for emissions considerations and to keep the drivetrain whole, the Turbo was available only with a three-speed automatic transmission. And even then the automatic had to be substantially strengthened to handle the turbocharged engine's torque, with larger input and output shafts and numerous internal changes.

Chassis changes were made as well to make sure the added power wouldn't launch drivers into the roadside scenery. There were larger wheels and tires (cast alloy 15x6 wheels with Bridgestone Potenza P205/60R-15 tires) matched with *softer* springs front and rear. The front anti-roll bar went up 1 mm to 23 mm while the rear bar remained at 20 mm. Shocks were recalibrated to give more rebound control, a little at the front and a lot at the rear, and less control of jounce. Stiffer bushings were used at the rear. Proving that stiffer suspension isn't always better for handling, the 280ZX Turbo handled better than a standard model. The power-assisted recirculating ball steering of the standard ZX was replaced by power assist rack-and-pinion in the Turbo, eliminating the complaint about numb steering.

Turbo models came well-equipped, with air conditioning, AM/FM/stereo cassette, headlight washers, cruise control, and electric windows as standard equipment. Changes to the instrument panel included dropping the ammeter in favor of a boost gauge (a warning light took the ammeter's place), and an oil temperature gauge shared space with the oil pressure gauge. Turbo models were identified by "TURBO" badges on the front fenders and hatch, twin exhaust outlets, and a functional hood scoop.

The list price reached an all-time high of $16,999, but testers regarded that neighborhood "a remarkable bargain." The Turbo's handling approached that of earlier Z cars and acceleration was good enough to better a Corvette 350 with a four-speed manual, turning the quarter-mile in 15.6 sec, compared to the Corvette's 16.0 sec. The Porsche 924S Turbo wasn't even close (although both Porsche and Corvette could out-corner the ZX's .754 lateral acceleration on the skid pad). The 280ZX Turbo, per *Road & Track*, had a "handsome interior, comfortable interior, gobs of performance and sporting handling characteristics," making it not only a bargain but also "a damned exciting car." Nissan scheduled 7,000 Turbo ZXs for the U.S. market in 1981, about one-tenth of total expected Z sales.

1982

Talking ZXs debuted in 1982. Leave a door ajar and a soft female voice would tell you so. The same for leaving the lights on, letting your fuel level get low, or leaving your parking brake engaged. It was most popular with stand-up comedians who liked to tell jokes about cars they couldn't afford.

There were other changes to the 1982 Datsun 280ZX however, particularly one that made the car better looking. Soft urethane body-color bumpers replaced the steel-faced bumpers of the year before, furnishing a more finished look. The grille was reworked to improve appearance and airflow, and body side moldings were added, coordinating with rub strips on the front and rear bumpers. The taillamps were changed too, while the NACA-type hood scoop introduced on the 1981 1/2 Turbo ZX was added to all 1982 ZXs.

Interior upgrades included better upholstery, remote rear side window openers on 2+2 models, and "suede-like accents" with the optional leather seating. An "ambiance" sound system had a reverb feature that, fortunately, could be overridden.

Functionally, there was good news for the enthusiast: rack-and-pinion power steering became standard equipment and a five-speed overdrive manual transmission became available with the Turbo in August. The Turbo also was given a longer 3.54:1 final drive ratio, compared to 3.90:1 for the naturally aspirated ZX, but the Turbo's five-speed was a Borg-Warner T5 instead of the Nissan unit, with gearing that made the two about equal. In testing, the auto-

The standard tire on the 1983 280ZX was enlarged to a 205/70HR-14. *Nissan*

matic actually proved faster, attributed to the need to lift off the throttle—and thus lose boost—when shifting the manual. *Road & Track* recommended the automatic with the turbo, and the five-speed with the naturally aspirated engine.

And finally, for fast guys with more than one friend, the turbo engine was available in the 2+2.

1983

Nissan continued to develop the 280ZX in 1983, offering new options that, for better or worse, tilted the car more toward luxury and less toward sports. A leather/digital option package combined the leather interior with "nu-suede" accents, digital instrumentation, automatic temperature control, bronze-tinted glass, electric defogging for the mirrors and an automatic defogger for the rear window, and a premium sound system, all for $990.

The digital gauge idea was then in its infancy, and the new Datsun instrument panel wasn't one of the better ones. A collection of rectangular readouts included a digital speedometer and a digital/analog tachometer that was all but unreadable and non-linear in its electronic analog portion.

Tire size for the naturally-aspirated Z was fattened to 205/70HR-14 while the Turbo stayed at 205/60R15. Nissan "commonized" suspension calibrations between turbos and non-turbos, making spring and shock absorber damping rates the same. As a result, *Motor Trend* criticized the '83 280ZX Turbo's suspension for not being the equal of the engine, particularly when pushed hard.

The 1983 Turbo-ZX in solid color (non-two tone) can be identified by what Nissan called a "blackout package" including, most notice-

1979 Datsun 280ZX

Base Price: $7,968

Major options: GL package, $2,284; California emissions, $105; automatic transmission, $295; two-tone metallic paint, $99

General

Layout	rwd
Curb weight (lb)	2,825
Wheelbase (in)	91.3
Track, front/rear (in)	54.5/54.3
Overall length (in)	174.0
Width (in)	66.5
Height (in)	51.0
Ground clearance (in)	5.9
Seats	2
Trunk space (cu-ft)	19.8
Fuel tank capacity (gal)	21.1

Chassis and Suspension

Frame type	Unit steel
Front suspension	MacPherson struts, lower lat eral arms, compliance struts, coil springs, anti-roll bar
Rear suspension	Semi-trailing arms, coil springs, tube shocks, anti-roll bar
Steering type	Rack & pinion
Steering ratio	19.6:1
Tire size	195/70HR-14
Wheel size (in)/type	14x5 1/2/steel disc

Brakes

Front type	Vented disc (vacuum assisted)
Size (in)	9.9 diameter
Rear type	Disc (vacuum assisted)
Size (in)	10.6 diameter

Engine

Type	SOHC Inline 6
Bore & stroke (mm)	86.0 x 79.0
Displacement (cc)	2,753
Compression ratio	8.3:1
BHP @ rpm	135 @ 5,200
Torque @ rpm (ft-lb)	144 @ 4,400
Fuel supply	Bosch L-Jetronic
Fuel required	Regular

Drivetrain

Transmission speeds/types	5/man and 3/auto
Gear ratios (:1)	
1st	3.32
2nd	2.08
3rd	1.31
4th	1.00
5th	0.86
Final drive ratio	3.36:1

Performance (Source: Road & Track, 11/78, 5-speed manual)

0-60 mph (sec)	9.2
1/4-mile (sec @ mph)	17.2
Top speed (mph)	121
Fuel economy (mpg)	22.0 observed

12+2 available.

ably, a "turbo" identification stripe on the side of the body.

Prices continued their climb. List price for the 280ZX GL was $14,799. The fully equipped 280ZX Turbo listed for $17,299.

It was clearly the end of the line for the second generation Z-car, however. Sales had been in a downward slide since 1980, falling from that year's 71,533 to 1982's total of 57,260. The profile, no matter how successful in 1970 or in years following, had become old. It was time for something new.

What to Look For

The 280ZX has a rather unfortunate reputation among sports car enthusiasts. It isn't, after all, a sports car. The maker intended it as sports-grand touring car, and that's how it should be judged. A little too much disco? Well, that's what the fashion was then. Given enough time, disco will be nostalgia rather than just painfully passé. The 280ZX will doubtless go through a similar evolution.

It should be noted that, despite the tape stripes and two-tone paint scheme, the 280ZX was no false front. The Turbo was faster than a contemporary Corvette 350, clocking a 15.6 in the quarter—still respectable today—almost a half-second ahead of the 'Vette.

In the meantime, the ZX is exceedingly affordable for those who like it. On the other hand, there are still enough "used cars" around to make selecting a good one a lengthier task. There may be more wheat, but there's more chaff, too.

Rust isn't quite the problem on these cars as it was with the 240-260-280 generation of Zs, but a careful examination is still a good idea. At the going prices and with availability of cars, a rusty 280ZX isn't worth the trouble or (especially) the expense.

1981 1/2 Datsun 280ZX Turbo	
Base Price	$16,999
General	
Layout	rwd
Curb weight (lb)	2,995
Wheelbase (in)	91.3
Track, front/rear (in)	54.9/54.7
Overall length (in)	174.0
Width (in)	66.5
Height (in)	51.0
Ground clearance (in)	n.a.
Seats	2
Trunk space (cu-ft)	19.8
Fuel tank capacity (gal)	21.1
Chassis and Suspension	
Frame type	unit steel
Front suspension	MacPherson struts, lower lateral arms, compliance struts, coil springs, anti-roll bar
Rear suspension	Semi-trailing arms, coil springs, tube shocks, anti-roll bar
Steering type	Rack & pinion, power-assisted
Steering ratio	n.a.
Tire size	P205/60R-15
Wheel size (in)/type	15x6 in/cast alloy
Brakes	
Front type	Vented disc (vacuum assisted)
Size (in)	9.9 diameter
Rear type	Disc (vacuum assisted)
Size (in)	10.6 diameter
Engine	
Type	SOHC turbo Inline 6
Bore & stroke (mm)	86.0x79.0
Displacement (cc)	2,753
Compression ratio	7.4:1
BHP @ rpm	180 @ 5,600
Torque @ rpm (ft-lb)	203 @ 2,800
Fuel supply	Bosch L-jetronic
Fuel required	Regular unleaded
Drivetrain	
Transmission speeds/types	3/automatic
Gear ratios (:1)	
1st	2.46
2nd	1.46
3rd	1.00
Final drive ratio	3.55:1
Performance (Source: Road & Track, May 1981)	
0-60 mph (sec)	7.4
1/4-mile (sec @ mph)	15.6
Top speed (mph)	129
Fuel economy (mpg)	20.0 estimated

Chapter 10

1990-1996 Nissan 300ZX

★★★	1984-1989 300ZX
★★★1/2	1984-1989 300ZX Turbo
★★★★	1984 300ZX 50th Anniversary
★★★★	1988 300ZX Limited Edition

The look was almost entirely new. Gone were the sugar-scoop headlamp bezels and the power bulge on the hood that looked like a boat hull upside down. The fastback remained, although that was hardly unique to any sports car. The fact was that the 1984 Nissan 300ZX was an almost completely new car: A wedge-shaped profile with semi-concealed pop-up headlamps, a turbo and non-turbo 3.0-liter V-6 engine, a new electronically-controlled four-speed automatic and revised five-speed manual transmissions, and revised suspension and bigger brakes, not to mention new interior features.

It was a new car with new power and handling, but with more luxury than ever it was continuing a trend that dated back at least to the 260Z 2+2. Z-cars became more luxurious every year, more grand-tourer and less pure sports. But sales would climb to new records, so from that standpoint, Nissan was doing the right thing.

The styling abandoned the trademarks established by the original 240Z, the traditional shape giving way to a Guigiaro-influenced wedge. The radiator opening was under the bumper and pop-up dual-lens headlamps fit the angular, hard-edged shape that swept back

The Goertz-derived rounded Z-car styling gave way to a trendy Guigiaro-influenced wedge in 1984. The base model 300ZX is shown here. *Nissan*

The 2+2 was continued through the model change. A 300ZX 2+2 GS is shown here. *Nissan*

to a sharply truncated tail. The rear quarter-windows were straight-edged triangles and the backlight was rectangular and almost flat, though not horizontal. Turbo spotters noted the raised scoop (not popular with all critics) on the driver's side of the hood and a sole "TURBO" badge on the tail. The new body dropped the drag coefficient (Cd) from 0.36 for the 280ZX to 0.31 for the standard 300ZX and 0.30 for the Turbo, the latter benefiting from a front chin spoiler and small rear wing.

The new style's low hood was no doubt facilitated by the wholly fresh 60-degree V-6 engine. The engine, designated VG30E, was lower than the old inline-six and about 10 in shorter front to rear. Only slightly wider, it weighed 39 lb less than the 280ZX engine. To reduce vibration and engine noise level, Nissan used finite-element analysis to design the thin-wall cast-iron block. A main bearing cap in a single casting connected the four main bearings and made the block stronger, making a very strong bottom end for the engine. The connecting rods were forged steel and the even-firing crank was iron.

Aluminum heads had a single overhead camshaft apiece, and each was cogged-belt-driven and operating two valves per cylinder with cast aluminum rocker arms and hydraulic lifters. The spark plug was centrally located in a pentroof-design combustion chamber with a large squish area for reduced emissions and improved anti-knock characteristics.

The 300ZX was available from the start in naturally aspirated and turbocharged versions. The former was rated at 160 bhp and 173 ft-lb of torque at 4,000 rpm. Adding a Garrett AiResearch T-5 turbocharger (with 6.7 psi boost, with a knock sensor and compression dropping from 9.0 to 7.8:1) brought power to a peak of 200 hp at 5,200 rpm, with redline at 6,000 rpm. The turbo motor's torque topped at 227 ft-lb at 3,600 rpm. Ignition and fuel injection,

The 1984 300ZX Turbo was easily spotted with it's raised—and controversial—hood scoop. *Nissan*

The 1984 300ZX Turbo was the most complicated yet, but it retained semi-trailing arm rear suspension. *Nissan*

with a Hitachi hot-wire airflow meter, were computer-controlled.

A choice of five-speed manual or a three-speed or four-speed overdrive automatic was offered with the standard or turbo engine. The turbo's five-speed was a beefed-up version of a Borg-Warner unit; the non-turbo had a Nissan five-speed. The automatics were all new and once again two different boxes were used with the two different engines. The transmissions had two shift schedules, one for performance and one for economy. The non-turbo's box decided electronically which schedule to follow based on vehicle speed and the rate and degree of throttle movement. The driver chose the program with the Turbo's transmission. The automatic in the naturally aspirated engine's transmission had an automatic lock-up torque converter for improved fuel economy.

Suspension was the least-changed part of the Z-car, still with MacPherson struts at the front and semi-trailing arms at the rear. The geometry of the struts, however, was altered, with two degrees more caster and low trail with the kingpin axes behind, rather than in line with the wheel center, all for better stability at high (100 mph+) speeds. More anti-dive came from lowering the mounting point of the tension rod as well as the greater caster; shortened transverse links reduced bump steer. New bushings in the front suspension

The 3-liter turbocharged V-6 produced a healthy 200 hp in 1984. *Nissan*

The rear suspension of the 300ZX was mounted on a subframe. *Nissan*

contributed more to a quieter ride than precise handling.

The rear suspension was new in that the shock absorbers and coil springs were separated, the arrangement taking less room from the cargo area. Larger bushings on the rear suspension plus dual-path bushings reduced "thump." The stiffer suspension reduced the squat that even Nissan admits had characterized the 280ZX under acceleration. A 22.2-mm anti-roll bar at the front was matched with a 22-mm bar at the rear, with Nissan claiming greater roll understeer resulted in better straight-line stability.

Standard on the turbocharged ZX were DeCarbon-type shock absorbers with driver-adjustable firmness. Unlike other systems, which used solenoids on the shocks, the 300ZX system was novel, containing an electric motor within each shock tube. The motor rotated a shutter that selectively aligned different openings for different resistance. The cockpit switch was marked "firm," "medium," and "soft," but testers claimed little difference in the ride—except for speeding over potholes—and even less change in handling.

Overall, however, the suspension had a wider track, added caster at front, and 2 degrees negative camber at rear, more linear toe change front and rear during suspension travel, and improved anti-dive and anti-squat. It added up to easier straight-line driving and more predictable transient handling. Power rack-and-pinion steering from ZF of Germany had assist that faded with speed for better road feel while still helping in parking maneuvers.

Wheel size was 15x6 1/2 in on all models, but Turbos had five-bolt hubs and P215/60Rx15 Goodyear Eagle GTs while non-turbos had four lugs and came with either Toyo or

Black moldings like those seen here further distinguished Turbo models from non-turbo-equipped 300ZXs in 1985. *Nissan*

Bridgestone tires. Interestingly, standard and Turbo wheels shared the same design, differing only in the number of lugs. Brakes were the same on both models: power-assisted, ventilated 10.8-in front brakes and 11.4-in solid rear discs, with twin vacuum servos to help fend off brake fade.

The 300ZX still catered to the 2+2 crowd, but only with the standard engine. The Turbo came in two-seater trim only. Still, despite adding 7.8 in to the overall length and 7.9 in to the wheelbase, Nissan wanted to make it harder to tell a 2+2 from the coupe. The lines were shaped more subtly and the longer rear section lightened visually with larger side quarter-windows. The 2+2 was still a snug fit in the rear and not a place normal adults would want to inhabit for a cross-country cruise.

The front left seat was the best place to be. The dash continued the tradition of supplemental gauges in the center of the dash, but these were down to two with a traditional hooded nacelle over the main instrument cluster in front of the driver. Base ZXs came with easy-to-read white-on-black analog gauges, a 145-mph speedometer on the left, an 8,000-rpm tachometer on the right, flanked by fuel and engine temperature gauges. Oil pressure/temperature and turbo boost were canted toward the driver in the holes on the center of the dash, with an electronic diagnostic display immediately below that warned of lamp failures and doors ajar, plus a dual trip odometer.

The leather/digital instrument option—in addition to leather seats—included a new main instrument cluster with a digital speedometer and vertical electronic bar graphs for fuel, engine temperature, oil pressure, and voltage. Centered was a graphic display of the engine's power curve with illuminated bars that displayed the engine's rpm. Strangely, the markings were logarithmic rather than linear—and if that weren't enough, it was supplemented by a digital tachometer readout. Critics found it all hard to read and dimly lit, preferring the standard gauges. The digital instrumentation package also put an electronic acceleration/braking display and instantaneous fuel economy readouts in the left-dash center hole and an electronic compass in the right.

The leather/digital instrument option was revised for the 1984 model, the tachometer was horizontal and logarithmic, and it was imprinted with a simulated power band. *Nissan*

"Bodysonic" was the name Nissan gave to an audio system enhancement that enhanced the feel of the lower octaves via speakers in the seat bottoms. Hey, Mr. Bassman!

A new steering wheel had spokes at an un-sports carlike 5 and 7 o'clock, but at least the vinyl cover had thumb grips just below 3 and 9 o'clock. The interior overall was more luxurious than ever, with better color coordination and such, but there were irritating details such as imitation stitching in the dashboard vinyl.

All in all, performance was up (top speed of the Turbo was electronically limited to 136 mph), cornering grip was improved, and the car was in almost every way improved over the 280Z, the fastest Z-car yet. But testers found the new 300ZX too heavy, too soft, and too luxury laden to be a true sports car. It was, though, a better than ever grand tourer.

A run of 5,000 limited-edition 50th Anniversary 300ZX Turbos was built to commemorate the formation of the Nissan Motor Company from the DAT Jidosha Seizo Company in 1935. Although the drivetrain was standard Turbo fare, Nissan specified two-tone paint in metallic silver, along with "ground effects" rocker-panel appliqués, fender flares, and "turbo-finned" 16-in wheels with Pirelli P7s, to make the Anniversary model stand out even

All 300ZX models came with small black spoiler. *Nissan*

among other ZXs. The silver treatment also included the T-top. Springs stiffer (by about 10 percent) and a front anti-roll bar 1.0 mm thicker than usual combined with a stiffer recalibration of the cockpit adjustable shocks to tighten up a notch, although critics said not enough. The full slew of options came on the special, bringing its price to a company high of $25,999.

Base price for a two-seat non-turbo 300ZX—the least expensive model possible with no options—was $15,799, but it hardly made for a Dickens Orphanmobile. Standard equipment included power steering, brakes, windows and mirrors; air conditioning; full instrumentation; eight-way adjustable seats; cruise control; rear wiper and defroster; tilt steering; alloy wheels and big radials; and a 40-watt four-speaker AM/FM/cassette stereo. Add the T-top for $800 more, and the automatic transmission cost $420. Pay $1,900 for the leather/digital instrument package and you also got climate control. The Turbo started at $18,199.

You were lucky, though, if one could be found at list. Dealers were asking thousands over list and getting it. Nissan had buyers standing in line no matter what the string-back-glove set said. Sales for calendar year 1984 would set a new record for the Z-car, hitting a grand total of 73,101 units. The ZX continued as the most popular sports car in America.

1985

Changes the first year after a new model introduction are usually what a maker calls "refinement." There were no mechanical changes, although the Leather Trim option was separated from the Electronic Equipment Package (digital instrumentation; trip computer; AM/FM/cassette stereo with nine-band graphic equalizer, eight speakers and fader/balance controls; automatic temperature control; defogging outside mirrors; and steering wheel-mounted audio system and cruise control switches).

The T-roof was made standard equipment and included new lockable panels. Otherwise, new were two-tone color combinations: black with gold lower body or dark metallic blue with lower pewter body color. Two-tone models had body-colored, rather than black, bumpers. Black moldings and smoked rear lights and reflectors distinguished Turbo models.

Prices continued to climb. The two-passenger Coupe listed for $17,199, while the 2+2 went for $18,399 and the Turbo—sans options—had $19,699 on the window sticker. Sales continued strong, at 67,409 units, for the model's second year.

1986

The third year after a new model intro is usually time to "refresh" it. So it went with the 1986 300ZX. Refresh meant in this case integral fender flares adding about 2 in on either side to cover the 7.0x16-in wheels with 225/50VR-16 Bridgestone Potenza tires that came with the Turbo. (The standard model made do with 6.5x15-in wheels and sixty-series tires.) "Ground effects" rocker panel extensions and chin spoilers became standard, and Turbo models lost the controversial hood scoop. The federally mandated, high-mounted center stop light made its debut.

One tester noted that, contrary to earlier reports of little difference between the adjustable shock absorber settings, the 300ZX's system seemed "the most crisply staged from one setting to the next." The four-speed automatic was still optional but revised. A ceramic impeller for the turbocharger didn't increase performance but Nissan said the durability was improved.

Inside, a sportier four-spoke steering wheel was good news, with (when ordered) steering wheel-mounted sound system controls. Seats were lowered an inch to better fit tall Americans. And after being standard equipment for a single year, the T-roof became optional again for 1986.

Prices were up again, a no-option 300ZX topping 20k for the first time, at $20,599. The base model checked in at $17,199, the 2+2 at $18,399. Sales slipped to 52,936, the normal trend as a vehicle ages, but the 300ZX was also stiff competition from the Toyota Supra, Mazda RX-7, and a virtual host of others.

1987

The styling of the 300ZX was tidied significantly for 1987, with the hood, air dam, and bumper integrated for better aerodynamics, with driving lights moved under the front bumper. At the rear there was a new full-width taillamp, while new wheels with a brushed alloy finish denoted the non-turbo 300ZX; the Turbo had wheels with a charcoal finish. A new turbocharger was fitted but with no effect on rated power output, while fuel-injector modifications bequeathed a smoother idle.

Better handling was the result of a major reworking of the ZX's steering and suspension. Springs and bushings were upgraded in front and the anti-roll bars and shock absorber rates were recalibrated for better cornering and response, also helped by changes in the power steering pump. Larger brake calipers improved fade resistance.

Prices moved up again, the dollar still sinking against the yen. For 1987, the base 300ZX listed for $18,499, the 2+2 for $20,649, and the Turbo $21,399. Despite the improvements, sales dropped precipitously to 33,649, less than half of only a few years earlier, the combined result of a tightening market and a growing list of competitors.

1988

Nissan engineers managed to find another five hp both for turbo (compression raised from 7.8:1 to 8.3:1) and non-turbo engines. The new "High Flow" turbocharger reduced turbo lag for better mid-range response. A bright silver finish for the Turbo's wheels ended complaints about the prior year's gray wheels that "always looked dirty."

The Nissan 300ZX Limited Edition was the fastest Japanese car in America—with its speed governor disconnected. Although usually limited to 136 mph, the 300ZX LE went 153 mph in a *Motor Trend* top speed test. The LE had shorter springs for a lower stance and an improved "European" front air dam.

Significant increases marked 300ZX prices for the year. The GS with no options was lowest at $21,699, the 2+2 GS next at $22,849, the Turbo at $24,099, and the LE Turbo tops at a base price of $25,099. The Turbo represented 18.0 percent of 300ZX sales for 1988, down to 19,357 overall.

1989

It was the final year for the third-generation ZX, and with the release of four all-new models and with an all-new Z-car due in the spring, the 1989 300ZX was a carry-over model. Prices climbed to $24,649 for the 300ZX GS and $25,949 for the Turbo.

Automotive journalists, for whom familiarity usually breeds contempt and who had equivocated over the luxury/sports split personality from the beginning, were turning on the model with a vengeance. In its October 1988 issue, *Automobile* praised the engine's power—"seamless, free from turbo lag, and responsible

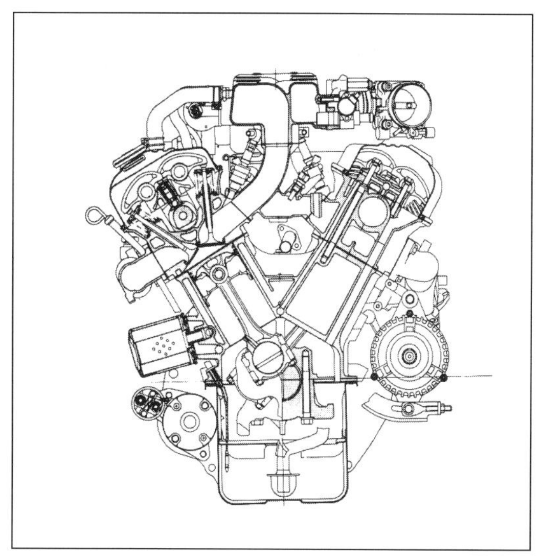

Here is a cross-section of the Nissan 300ZX V-6 engine, showing valve actuation by the single overhead camshaft. *Nissan*

A styling freshening in 1986 included wider fender flares, prominent rocker panels, and new wheels. Also, the turbo lost its hood scoop. *Car and Driver*

for low-seven second 0-to-60 squirts"—but suggested it was 500 lb overweight and criticized "too much on-center numbness and wander" and, near cornering limits, its "habit of heeling over and feeling wishy-washy and disconnected, giving the impression that the bushings have turned to bubble gum . . . understeer appears in spades, except for those brief and amazing flashes of lift-throttle oversteer. . . . " and so on. Conclusion: They'd wait for the new one.

Which, as it turns out, would hit the streets

This 1986 Turbo is equipped with the leather and digital instrument options. There's even a "TURBO" logo on the front face of the seats. *Car and Driver*

The 300ZX was smoothed and had its driving lights relocated to below front bumper in 1987; a 1988 model is shown. *Nissan*

in the spring. Records overlap, but monthly sales of the 300ZX were hovering in the vicinity of 1,300 per month—until April when they jumped to 3,562. The new guy must have come to town.

What to Look For

The shopper for a 1984-1986 300ZX should have a wide selection from which to choose. The later cars are of course the better ones, prettier, better handling, and more powerful, but being newer they'll be higher priced and, since fewer were made, they'll be harder to find. The classifieds of a big-city newspaper should have a number of listings and even the more reputable used car lots may have them in inventory. At this point, third-generation cars are still "used cars" and should be expected to drop in value regardless of relative desirability. They certainly aren't as sleek and modern as the fourth-generation cars, but they're much less expensive and definitely better than driving a sedan.

Model: 1984 Nissan 300-ZX Turbo
Base price: $23,360

General

Layout	RWD
Curb weight (lb)	3,080
Wheelbase (in)	91.3
Track, front/rear (in)	55.7/56.5
Overall length (in)	170.7
Width (in)	67.9
Height (in)	51.0
Ground clearance (in)	n.a.
Seats	2
Trunk space (cu-ft)	18.5
Fuel tank capacity (gal)	19.0

Chassis and Suspension

Frame type	Unit steel
Front suspension	MacPherson struts, lower lateral arms, compliance struts, coil springs, anti-roll bars
Rear suspension tube shocks,	Semi-trailing arms, coil springs, anti-roll bar
Steering type	Rack & pinion, power-assisted
Steering ratio	n.a.
Tire size	P215/60R-15
Wheel size (in)/type	15x6 1/2/cast alloy

Brakes

Front type	Vented disc
Size (in)	10.8 diameter
Rear type	Disc
Size (in)	11.4 diameter

Engine

Type	SOHC turbo V-6
Bore & stroke (mm)	87.0x83.0
Displacement (cc)	2,960
Compression ratio	7.8:1
BHP @ rpm	200 @ 5,200
Torque @ rpm (ft-lb)	227 @ 3,600
Fuel supply	Bosch L-Jetronic
Fuel required	Regular unleaded

Drivetrain

Transmission speeds/types	5/man	4/auto
Ratios (:1)		
1st	3.35	n.a.
2nd	2.05	
3rd	1.37	
4th	1.00	
5th	0.77	
Final drive ratio	3.54	

Performance (Source: Road & Track, 1/84)

0-60 mph (sec)	7.4
1/4-mile (sec @ mph)	15.7 @ 86.0
Top speed (mph)	133
Fuel economy (mpg)	17.0 observed

Chapter 11

★★	1990-1996 300ZX
★★1/2	1990-1996 300ZX Turbo
★★★	1995 SMZ

1990-1996 Nissan 300ZX

"We want the 300ZX to be the world's number-one sports car." So said Nissan's executives upon introducing the all-new 1990 300ZX. Never mind the ambiguities of such a statement—Sales? Performance? Number one in what? It was an ambitious goal nonetheless. More important, however, was what wasn't said. This new car wasn't about sports-luxury or luxury-sports. It wasn't even grand touring. It was a *sports car*.

Even more, with a completely new contour, it looked the part. Its 2,960 cc V-6 had double-overhead cams and a 10.5:1 compression ratio and made 222 bhp, so it went like a sports car. The suspension was new, too, with front and rear suspension more like a race car's than the old ZX's, so it handled like a sports car. Monster brakes, wide tires, and 16-inch wheels completed the package. And although there was a 2+2, Nissan was indeed reaching for that brass ring.

The 1990 300ZX was as stunning as a concept car turned loose on the highway. *Nissan*

The shape of the 1990 300ZX struck one first. "No more poity-toity, hippy-dippy fenderware for this debutante," said *Car and Driver*. "Her very appearance hurls a silent come-on. She winks, 'Let's dance!'" The design goals of a forward cabin, short overhangs, flush wheel-to-body relationships, extended wheelbase, short hood, and taut and fluid shape were clearly met, and it was stunning. In profile, front-quarter view, or coming at you for the first time on the highway, the new 300ZX was enough to make you leave whatever else you were driving on the side of the road.

"Gimmick-free" was the password for the styling, which made few concessions to aerodynamics but still had a coefficient of drag of 0.31—its angular ancestor was actually slightly better in the wind tunnel—and looked wholly like a sports car. The double-slot front end met aero and styling needs, while the incredibly inclined (60 degrees) headlamps with projector-type low beams gave the 300ZX a distinctive face. The windshield was laid back as well and there was a slight "Coke-bottle" flow to the rocker panel. For 1990, a tinted glass T-top was standard and, true to tradition, the fastback backlight was part of a hatchback, but one that completely opened the rear of the car above its rear character line. Taillamps were contained in a black panel above the integrated rear bumper.

It was more than just styling, of course. The chassis underneath was entirely new and profoundly more rigid that its predecessor's, partly to meet the requirements of its new front and rear suspension. After twenty years, the MacPherson struts at the front were history. But rather than a simple me-too double A-arm arrangement, an altogether new front suspension was devised. A lower arm and a tension rod formed the lower half, while the upper arm connected to the chassis higher than the top of the wheel. The arm was articulated and carried a strut/spring just inboard of the wheel end. It looked impossible, but Nissan's Cray computers said it would keep the front wheels vertical to the pavement anywhere this side of a roll-over.

The rear suspension was equally complex and was also developed on the Cray. Borrowed from the acclaimed rear suspension of 1988's new 240SX, each rear wheel's suspension geometry was controlled by five control arms. The system, debuting on Nissan's MID-4 show/concept car in 1985, comprised a lower A-arm, a rear control link, and two upper links. The upper links were essentially a "split" upper A-arm, positioned to provide added stability in turns.

The 3.0-liter V-6 engine was the same as that in the previous 300ZX—at least as far as bore and stroke. Everything else, though, was changed. The block was the same, but redesigned, along with the crankshaft and connecting rods.

The cylinder heads were completely new, with two camshafts per head and four valves per cylinder for greatly improved volumetric efficiency. With the four valves in a pentroof arrangement, the spark plug could be located centrally for better combustion. A combustion ratio of 10.5:1 mandated premium fuel but paid off in full with higher output.

The intake system had a single airflow meter, but downstream was independent for each bank, even with two throttles. Long and large intake ports narrowed about halfway in to speed the velocity of the intake charge. Fuel delivery was by port fuel injection.

The Nissan Valve Timing Control System (NVCS) was the company's version of variable valve timing. It worked by not attaching the intake cam's pulley directly to the shaft, but rather placing a helical gearset in between. This hydraulic-mechanical device allows rotation of the intake cam, altering valve overlap for smooth and efficient idling *and* maximum power at higher rpm. Lifters were hydraulic, and with the overall design of the valvetrain and intake system, good for the engine's 7,000 rpm redline.

The new interior was a stylist's fantasy, with twin cockpits divided by an organically contoured console. *Nissan*

There was no distributor on the 1990 300ZX. Instead the Nissan Direct Ignition System (NDIS) had a compact coil on each spark plug fired by a trigger signal from the Electronic Concentrated Control System (ECCS—Nissan had more acronyms than the U.S. Navy), basing the precise firing instant on data provided by a crank-angle sensor. No whippy distributor shaft meant more accurate spark timing, and no high-tension wires meant no loss of ignition energy at high rpm.

A forged crankshaft replaced the cast crank of earlier models, and an oil jet squirted at the bottom of the piston for increased cooling. The result of the engine engineering was a naturally aspirated engine that made more horsepower than the earlier turbo motor. Output was rated at 222 bhp at 6,400 rpm, and 198 ft-lb of torque at 4,800 rpm. Zero-to-sixty came

The profile of the 1990 Nissan 300ZX featured some of the most radically laid-back headlamps in the industry. *Nissan*

in the 6 sec range, with the quarter-mile about 15 sec and top speed somewhere in the neighborhood of 150 mph, depending on the neighborhood and the transmission.

A five-speed manual transmission was standard equipment, and Nissan redesigned the shifter. Instead of the previous arrangement—which, with the lever mounted directly to the transmission case, angled the lever back toward the driver—the new shift lever was mounted to the floor and was connected to the transmission via a support rod. The vertical shifter made changing gears easier. The new design also allowed a shorter—and more rigid—transmission case, and hence reduced NVH. Another change was the use of double-cone synchronizers on second and third gears to reduce shift effort and, as a result, reduce lever throws by more than an inch. The final spiff on shift feel was a shift knob 180 grams heavier than the previous Z's.

Prototype testing yielded another needed change in the five-speed: The new 300ZX's ability for prolonged acceleration in the range of 0.8 g starved the fourth-gear counterdrive of lubrication, requiring longitudinal baffling to keep the gear from seizing.

The optional four-speed electronic transmission had torque-converter lock-up for improved highway fuel mileage, but was remarkable for the use of electronics in its operation, particularly the transmission's "notifying" the ECCS engine-management system of impending shifts. The ECCS momentarily retarded ignition timing to reduce torque for smoother upshifts.

At the rear, the limited-slip differential used high-viscosity silicon oil to transmit torque between the axles. When one wheel began to slip, the differences in rotational speed caused sheer in the silicon oil, which then thickened and shifted more torque to the wheel with traction.

Lightweight five-spoke aluminum wheels were standard. The 16 x 7 1/2-in wheels had a very open design, both for brake cooling and to show off the big calipers. Low-profile P225/50VR-16 steelbelted radials, either Michelin XGT-V or Dunlop D40 M2, were normally fit.

That much tire and a car the size and weight of the 300ZX required power steering, preferably speed-sensitive to allow assist for parking while retaining road feel at speed. Conventional speed-sensitive systems could be caught out with low-assist pressure at speed, however, so Nissan developed a new recipe that added complexity to the hydraulics but smoothness to its operation. Rack-and-pinion steering provided precision.

Nissan put big vented disc brakes at all four corners. The discs, 11.0 in diameter in front and 11.7 in diameter in the rear, were designed for maximum heat transfer with minimal distortion. Opposed-piston type calipers were used, and the calipers themselves were

The 2+2 was subtly different from the two-seater. Note the slightly larger rear side window and wider panel between door and rear wheel arch. 1991 model shown *Nissan*

made of aluminum for reduced unsprung weight (with the outside styled as they would be on display). They were about 4.0 to 4.5 lb lighter than comparable cast-iron designs would have been. The front caliper was a four-piston design, and the rear had two pistons per caliper. ABS (anti-lock braking system) was standard.

When the 1990 300ZX appeared in dealer showrooms on April 24, 1989, there were two models: the standard two-seater and a 2+2. You'd have a much harder time telling the new plus-two from its two-only sibling. External cues were few; unlike the earlier cars there was no break in the roofline to tattle on the 2+2. The rear-quarter windows were larger and met the lower window line at a more acute angle, and dedicated spotters would note that the fuel filler had been moved from in front of the left rear wheel to behind it. Otherwise, the slightly longer wheelbase (4.7 in longer) and overall length (8.5 in longer) and a bit more height (0.2 in more) was well disguised. Of course, though the seatback folded forward for extra luggage room, the rear seats provided little passenger room. Said *Car and Driver*, "Calling the Z a '2+2' is something of a stretch. You could toss some soft luggage (or unruly

High taillight placement accentuated the nose-down stance of the 1990 300ZX. *Nissan*

kids) in the back seats, but full-size humans would have to kneel."

Regardless of other seating arrangements, interior design focused on the driver, who was separated from the right front passenger by a massive console that swept down from the dash. Drivers would find controls easy to reach and use, with pods on either side of the three-spoked steering wheel. No vestige remained of the dash-center instrument nacelles, but digital instruments were gone, too. Only analog dials were available.

Prices were higher than ever, starting at $27,300 for the two-seater. But equipment was complete, and the option list short: the automatic transmission; an electronics package (Nissan-Bose audio system, automatic temperature control, power driver's seat, dual heated mirrors, and illuminated entry); and a leather package (which required the electronics package, and included leather seating, bronze tint glass and a cargo area cover).

Ecstatic magazine articles had appeared months before the 300ZX went on sale. Said William Jeanes in *Car and Driver:* "This swoopy, seductive supercar makes strangers honk and give a thumbs-up sign and causes small children to wobble on their bikes as they crane to gaze at Nissan's newest." *Road & Track*'s road test said, "If you think we ran out of superlatives in our introductory article on the 300ZX a short two months ago, think again. We've had our Webster's working overtime keeping up with our feelings about the new Z." Jim Miller in *Motor Trend:* "This latest iteration puts Nissan back into the fore in world-class sports cars."

1990 Turbo

Beginning with the April introduction of the 1990 300ZX, Nissan had promised a twin-turbo version, and with new model introduction in the fall of 1989, they delivered—and then some. If power was good, the 300ZX Turbo was an angel. The 1990 Nissan 300ZX Turbo was rated at 300 bhp at 6,400 rpm, with 3,600 ft-lb of torque at 3,600 rpm. It was time again for journalists to belly up to the thesaurus.

The whole rationale for the model, of course, was the turbocharging of the engine. But a V-6 with crossflow heads is not an easy engine to turbocharge. It's much easier, for example, to hang a turbo on an inline engine with the intake and exhaust on the same side. It shortens the plumbing. So to turbo a vee-engine, why not treat it as two in-lines? Rather than one large turbo, use two smaller turbos mounted on each exhaust manifold.

The 300ZX Turbo could be identified by grilles in the chin spoiler and the duck tail spoiler. *Nissan*

With less inertia each, they'll spool faster for less turbo lag.

So that's what Nissan did, almost. Complex plumbing was necessary to pull it off, however. From the turbos, hybrids of the Garrett T25 compressor housing with a T2 turbine, compressed air is routed to intercoolers located in each front corner of the Turbo Z and from there to the intake manifolds. The intake manifolds, however, instead feed the opposite bank of the vee with interleaved, long, straight runners (instead of tightly curved runners if—for no advantage—they went to the same side bank).

To optimize turbocharging, the engine's compression ratio was dropped to 8.5:1. Molycoating was used on the skirts of the cast-aluminum pistons, and for added cooling, a jet of oil sprayed into channels in the underside of the piston face. Special exhaust valves of Iconel (aircraft-grade steel) were used to endure higher exhaust temperatures, and stronger connecting rods handled the extra power. Connecting-rod bearings were changed to a high-strength copper/lead alloy called Kelmet. The Turbo was also fitted with an external cooler for the engine oil.

The turbos were both oil- and water-cooled, with the water cooling for the bearings designed to siphon until the turbo temp dropped below boiling.

To transmit the torque, the clutch and transmission were strengthened, with a sturdier second and fifth gear in the manual and more grip for the clutch (38 percent more capacity for the clutch plate). To keep pedal effort manageable, vacuum-assist was provided—and the clutch pedal was noticeably more difficult to depress with the engine off. The driveshaft also got bigger universal joints.

The four-speed automatic (with a lock-up torque converter) was also available in the Turbo, but in deference to the auto box's longevity, the engine's power output was cut back to 280 bhp by different camshafts and fuel injection maps. Torque remained the same as with the four-speed, however. With manual or automatic, the Turbo's limited slip differential carried a 3.692:1 final drive ratio, quite a bit taller than the non-turbo's gearing. Nissan felt compelled to govern top speed to 155 mph. The car could otherwise top 160 mph.

Turbocharging was limited to the two-seat version of the 300ZX and ABS was required. Front wheels and tires were the same as the base model, but the Turbo came with wider—245/45ZR16—radials on 1 in-wider rims (to 8 1/2 in). Nissan further enhanced the 300ZX's handling response with Super HICAS rear steering, a system (unlike competitors) that did not work at slow speeds, but prevented oversteer by turning (via computer control and hydraulic pump) the rear wheels into the turns. To prevent a "dead" feel in steering, a slight outward twitch to the rear wheels aided turn-in before the wheels were turned inward. Total steering angle is less than 1 degree in or out. Adjustable damping shock absorbers are controlled by a rocker switch on the dash, with a choice of Touring and Sport settings, the latter firmer. Nissan also put a quicker front steering ratio in the Turbo, 14.8:1 compared to 16.8:1.

The Nissan 300ZX Turbo could be distinguished from its lesser siblings by the slots in the front spoiler for the intercoolers and the spoiler on the rear deck, labeled "TWIN TURBO" for anyone who needed more help. Inside one could tell a Turbo by a different fabric for the upholstery or, more easily, the boost gauge wedged into the bottom of the 160 mph speedometer. Although home market ZX300s came standard with a steel roof, all U.S.-market cars for 1990 came with the T-bar roof.

Hardly surprising, the 300ZX Turbo met with rave reviews. The road test in *Road & Track* stated that "those accustomed to the thumb-twiddle/hold-on-for-dear-life thrill ride provided by the Porsche 911 Turbo will be disappointed; but those whose idea of a good time is tractable, predictable gobs of power will be pleased," and it concluded by claiming, "Five years ago, if you said the 300ZX was among the world's best-handling cars, you'd have been met with polite laughter. Say it now and you're simply stating a fact."

Said Csaba Csere in *Car and Driver*, "The only cars in the world that can better this performance are high-buck exotics." *Motor Trend* made it the magazine's 1990 Import Car of the Year and *Automobile* named it the 1990 Design of the Year, putting it on its "All

Nissan's 300ZX Convertible debuted in the 1993 model year. *Nissan*

Stars" from 1990 through 1994, an impressive feat considering the new competition every year.

The 300ZX Turbo was to no one's surprise the most expensive Nissan, but a "high-buck exotic" it wasn't. List for the twin-turbo sports car was $33,000, but that included everything important. The only extras were electronic climate control, a power seat, leather upholstery, and the automatic transmission. Prices for the lesser models were also up slightly for the fall season, with the non-turbo two-seater up to $27,900, and the 2+2 at $29,100. Sales totaled 22,183 for the year, with 26 percent of those turbo equipped and 40 percent with an automatic transmission.

1991

Complaints about the T-top—not that there was anything wrong with it, just that it was required—prompted Nissan to add a non-T-top 300ZX to the U.S. product lineup; it had been offered on the home market from the beginning. It was available, however, only with the naturally aspirated engine, the five-speed manual, and in two-seater configuration.

Nissan also made some previous optional equipment standard, including the Nissan-Bose audio system, dual outside heated mirrors, illuminated entry system with courtesy light fadeout, cargo area cover, seatback storage pockets, and automatic temperature control covered in the base price. Options were limited to the automatic transmission, a driver's side airbag, and a leather seating package that included a power driver's seat.

Spotters could pick out new Zs by the Nissan badge on its nose. Although journalists had been able to persuade Nissan to leave the badge off in 1990 (it had been on "long lead" press cars), Nissan had decided to put it on for '91 and all subsequent years.

Prices took a couple-thousand-dollar jump. The new non-T-bar two-seater had a base price of $28,175, while the T-bar two-seater hit $30,300. An unadorned 2+2 listed for $31,270 with the Turbo reaching $34,570. Add the automatic transmission for $935 and the airbag for $500. Sales of the 300ZX in 1991 totaled 16,833; 35 percent had turbos, 42 percent were automatics.

1992

No mechanical changes greeted the 1992 300ZX line for 1992. Nissan instead increased the number of standard equipment items, putting a driver's-side air bag on all Z cars and a power driver's seat in all T-bar models. New cloth seat upholstery was an across-the-board change, along with placing a leather-covered shift lever in all T-top models with an automatic transmission. Turbo models came equipped with Goodyear Gatorback tires. The only other change was a revision of power doorlock mechanisms.

The option list was topped by the automatic transmission and the leather seating package on T-top models. All models could get a compact disc (CD) player for the first time, and base non-T-bar models were allowed to opt for the Bose audio system.

Naturally, prices were up again, with the base model pushing thirty grand at $29,705, the naturally-aspirated T-top model listing for $31,815, the 2+2 for $33,090 and the Turbo for $36,610. Leather added $1,075 and the Bose sound another $700. Out of a total 11,320 Z cars sold in 1992, 19.5 percent were turbocharged and 34 percent were shifted automatically.

1993

The Z car went topless in 1993, the first time the factory had ever sold a fabric-topped Z in its almost-25-year history. Based on the naturally-aspirated two-seater, it was available with either the four-speed automatic or the five-speed manual and came standard with the leather interior and the Bose sound system. The only options were the automatic transmission and a Sony CD player.

Back in 1990 Costa Mesa, California's R.Straman Company had offered to convert your very own 300ZX to a convertible for $8,500, although reports were that dealers placed the bulk of orders (for resale). It was a tidy conversion by an experienced shop, retaining the "basket handle" both for chassis reinforcement and because of door shape.

For its in-house convertible, Nissan went to ASC to design the drop-top and related mechanism. The Michigan firm would then make the parts and ship them to Japan for assembly. Thirty-seven structural reinforcements were also part of the convertible kit—which like Straman's had the basket handle—which added 210 lb to the two-seater's curb weight.

The 300ZX's top was a fully manual affair. To lower it, one unlatched the header clamps and then released the hard-plastic boot behind the seats. The top then collapsed into the well and was covered by the boot, which closed flush with the rear deck. With practice it took about 30 sec to drop the top. A flip-up air deflector, said to have been added for the benefit of smokers, reduced buffeting for all.

The 300ZX lineup was otherwise unchanged for 1993, though the continuing slip of the dollar against the yen forced prices up once again to a point at which no car sold for less than $30,000. The base model was $555 over that yardmark, while the naturally aspirated T-top Z sold for $32,730. The 2+2 listed at $34,040, the Turbo at $37,655 and the new convertible at $37,145. The automatic transmission was a $950 option.

Nissan sold 9,801 of the 1993 model year Z cars. Of that, 1,470 (15 percent) were turbocharged, 3,911 (39.9 percent) had automatic transmissions, and 2,068 (21.1 percent) were convertibles.

1994

Again Nissan made no changes in the major specifications of the 300ZX, just a couple of optional/standard feature realignments and some safety/environmental changes. A passenger-side air bag was put in the dash and, as a result, the door-mounted passive belt system was replaced by new three-point manual belts mounted on the B-pillar. The air conditioning system was revised so that a non-CFC refrigerant could be used.

A new standard feature was a keyless remote entry system, and the leather interior option became mandatory in Turbo models with automatic transmissions. The Turbo would be distinguished by a new rear spoiler, actually more of a wing freestanding from the rear deck lid.

The Turbo and Convertible models would be distinguished by prices that broke $40,000 for the first time as the dollar lost more ground against the yen. The manufacturer's suggested retail price (MSRP) for the Turbo was $40,099, and for the Convertible it was $40,879. The base Z was listed at $33,699, the T-bar non-turbo for $35,179, and the 2+2 for $36,489.

According to Nissan, calendar sales for the 300ZX totaled 4,836 units. Per the *Automotive News 1995 Yearbook*, of the total 4,421 model

The spoiler on the 1994 Turbo was elevated off the rear deck. *Nissan*

year sales, 1,587 were equipped with automatic transmissions. There were 597 convertibles and 1,018 Turbos.

1995

So little was changed in the 1995 300ZX that Nissan chose to highlight adding four new colors—including (for Jimi Hendrix?) Deep Purple Metallic—and a new "body-color front bumper fascia" (the black band from parking light to parking light was gone) as the major alterations. Not highlighted was the deletion of the four-speaker Bose audio system from the Convertible, which came only with a two-speaker AM/FM stereo (non-Bose) system. The coupe was the least-expensive 300ZX at $35,419, and the T-roof coupe listed for $36,489. T-roofs were standard on the 2+2 and the Turbo, with MSRPs at $38,169 and $41,959, respectively, while the Convertible carried a price tag of $42,659.

A—how to put this?—factory-related release was the SMZ, a $55,000 300ZX as modified by Nissan racer Steve Millen's shops. It was a quasi-official celebration of the Z-car's 25th anniversary, something the Nissan had celebrated with a cross-country relay marathon of Z-cars but not with a commemorative edition. Nissan kept its hands legally off this one. We suspect lawyers had something to do with that. But the cars were sold through Nissan dealers and were covered by Nissan's full three-year/36,000-mile warranty, which is more than earlier Millen-modified 300XZs were accorded.

To make a Millen SMZ, a 300ZX Turbo had its boost limit bumped about 2 psi to 11.9, and with Millen's low-restriction induction system and aluminized 3 1/2-in exhaust aft of the catalyst, horsepower was bumped to 365 bhp. No internal engine changes were deemed necessary. The requisite wheel and tire swap saw the SMZ shod in 255/40ZR-17 (front) and 265/35ZR-18 (rear) Yokohama Advan Nexus rubber on five-spoke Yokohama wheels. The wheels were 9 in wide at the front, 1/2-in wider at the rear. Thicker, larger-diameter, cross-drilled discs and big calipers from Nissan's home-market Skyline GT-R supercar were used, and bigger front and rear anti-roll bars complement stiffer variable-rate springs all around that lower the car 3/4 in.

The interior received carbon-fiber overlays for the window switch plates and the stereo system surround, plus a shift knob of genuine imitation carbon-fiber. If the wheels and lowered stance weren't enough to identify the SMZ, Millen also replaced the stock rear wing, using the standard mounting locations, with one high enough that one looked under it when looking out the back. And of course there's the SMZ decals. Douglas Kott, writing in *Road & Track*, noted that at 13.9 sec and 101.7 mph in the quarter-mile, and with .91 g cornering force, it was "significantly better than the stock car's performance without ravishing your kidneys or unduly pounding your eardrums." Millen scheduled a run of 300 SMZs, and, said Kott, "By all indications, it's a hot collectible, one that should gather a bare minimum of dust."

1996

The 1996 300ZX was unchanged from the 1995 model, Nissan instead trumpeting the model's 25-year heritage and five years' residence on *Car and Driver*'s "Ten Best" list and *Automobile* magazine's "All Stars" roster. Prices stood at $37,439 for the base coupe, $38,879 for the coupe with a T-roof. The 2+2 listed for $40,189, the Turbo (available with T-roof only) for $43,979, and the Convertible (with standard leather) topped the line at $44,679. Shortly after introducing the new models, however, Nissan announced that 1996 would be the last for the Z-car. With sales down again for 1995 and a substantial redesign required to meet U.S. federal side impact regulations for 1997, the company decided the Z no longer made economic sense. It had been an exciting 25 years, but it all was coming to an end.

What to Look For

Sports cars should have manual transmissions, at least according to purists, yet of all the 300ZXs—surely a bona fide sports car—close to half have automatics. Hold your scorn. Despite its 20 fewer horsepower and 30 extra pounds, the 300ZX Turbo Automatic is actually quicker than its manually shifted equivalent. According to tests by *Car and Driver*, the Automatic beat the five-speed in 0-to-100 mph testing by 15.8 sec to 16.3 sec. The reason? "The torque converter launches the car strongly, and the nonstop power delivery keeps the turbo boosting without disruption. The rush barely tapers as the automatic rips up through the gears," explained the magazine.

Still, traditionalists win when the road gets curvy, the manual box allowing for easier control—plus it's so much fun to use. Either way, traditionalists can have their way by leaving the T-tops off when its raining. That way they could have as much weather protection as an old English roadster with the top up.

Nissan has found that an "unusual turbocharger noise" might occur during driving, caused by reduced oil supply to the right side turbocharger and consequent turbocharger damage. The turbo damage may result from a clogged oil inlet tube. Beginning with VIN JN1CZ24D4RX545041 (engine number VG30-749015) in September 1993, Nissan began installing a revised oil inlet tube. Problems on earlier cars can be averted by frequent oil changes. The new inlet tube should be installed any time the right turbocharger is serviced or replaced.

Nissan has noted problems with automatic transmission slipping, shifting improperly, or failing to move in "D." A repair is available. Manual transmissions can develop a "grind" between 4th and 5th as well as a pinging noise in 5th gear. Vehicles from model years '90 and '91 have also had problems with clutch slipping, but dealers have a fix.

Dealers should also have corrections for a number of door window problems, including rattle, creaking and popping noises; wind noise; molding folding under and discoloring; and power window failures and judders. The 300ZX has also had problems with water leaks onto the front seat and into the luggage compartment.

The '90 and '91 models have developed brake shimmy, which can be felt as steering wheel shake while braking. Dealers have a technical bulletin on this problem. Nissan also warns that replacement tires for all years, particularly winter tires, should have the same dimensions and speed rating as the standard tires, and the company cautions against the use of aftermarket wheels. Any aftermarket wheels, however, should be the same size and offset as the standard wheels, and should have adequate ventilation for the brakes and sufficient clearance for the brake calipers.

1990 Nissan 300ZX

Base Price: $27,300

General

Layout	rwd
Curb weight (lb)	3,219
Wheelbase (in)	96.5
Track, front/rear (in)	58.9/60.4
Overall length (in)	169.5
Width (in)	70.5
Height (in)	49.2
Ground clearance (in)	5.0
Seats	2
Trunk space (cu-ft)	10.0
Fuel tank capacity (gal)	18.7

Chassis and Suspension

Frame type	Unit steel
Front suspension	Angled upper arms, lower A-arms, "3rd links," coil springs, tube shocks, anti-roll bar
Rear suspension	Upper lateral links, angled upper links, mid-lateral links, lower angled A-arms, coil springs, tube shocks, anti-roll bar
Steering type	Rack & pinion, power assist
Steering ratio	16.8:1
Tire size	225/50VR-16
Wheel size/type	16x7 1/2/cast alloy

Brakes (Power assist, ABS standard)

Front type	Vented disc
Size (in)	11.0
Rear type	Vented disc
Size (in)	11.7

Engine

Type	DOHC 32-valve V-6
Bore & stroke (mm)	87.0 x 83.0
Displacement (cc)	2,960
Compression ratio	10.1:1
BHP @ rpm	222 @ 6,400
Torque @ rpm (ft-lb)	198 @ 4,800
Fuel supply	Multi-point EFI
Fuel required	Unleaded premium

Drivetrain

Transmission speeds/types	5man, 4auto
Ratios, 5-speed (:1)	
1st	3.21
2nd	1.93
3rd	1.30
4th	1.00
5th	0.75
Final drive ratio	4.08

Performance (Source: Road & Track, 5/89)

0-60 mph (sec)	7.1
1/4-mile (sec @ mph)	15.5 @ 90.5
Top speed (mph)	148
Fuel economy (mpg)	23.5 (estimated)

2+2 available

separate box

1990 Nissan 300ZX Turbo (as 300ZX except as below)

Base Price: $33,000

General

Curb weight (lb)	3,501
Track, front/rear (in)	58.9/61.2
Seats	2

Chassis and Suspension

Steering type	Rack & pinion, power assist
Steering ratio	14.8
Tire size	225/50ZR-16 f / 245/45ZR-16 r
Wheel size (in)/type	16x7 1/2 front; 16x8 1/2 rear/cast alloy

Engine

Type	DOHC 32-valve turbo V-6
Compression ratio	8.5:1
BHP @ rpm	300 @ 6,400
Torque @ rpm (ft-lb)	283 @ 3,600

Drivetrain

Transmission speeds/types	5 man/4 auto
Gear ratios [5-speed] (:1)	
1st	3.214
2nd	1.925
3rd	1.302
4th	1.000
5th	0.752
Final drive ratio	3.692

Performance (Source: Road & Track, 12/89)

0-60 mph (sec)	6.5
1/4-mile (sec @ mph)	15.0 @ 96.0
Top speed (mph)	155 (Car and Driver, 2/90)
Fuel economy (mpg)	18.5 estimated

1977-1979 Datsun 200-SX

Datsun sedan fans waited. Since the demise of the 510 there hadn't been much to cheer for. The 610 had a bigger engine, the styling wasn't bad, and the chassis was largely the same as the 510's, but the luxury appointments made the car heavier and emissions rules took more away than the extra displacement added. The 710 not only gave up the 510's independent rear suspension, but it had that peculiar Buick-like bow wave character line on its side. Beautiful it wasn't.

Rumors circulated, however, about a new coupe on the Datsun 210 chassis but with the "big" L20B four—a 2.0-liter engine in a chassis

Tape striping on this 1978 200SX de-emphasizes the peculiar bodyside contour. Note the fully retracting rear side window. *Automobile Quarterly Photo and Research Library*

usually powered by a 1200 cc engine. Shades of the GTO. Even if the chassis weren't very sophisticated, it ought to be a rocket, at least relatively speaking. Oh, the anticipation.

Enthusiasm waned when photos began to leak, however. It was, well, odd looking. Perhaps the awkward 5 mph bumpers, throwing the front end out of balance, weren't completely Nissan's fault. And the peculiar C-pillar/backlight might have been able to work. But Nissan put another one of those silly bow waves on the 200-SX and lopped the tops of the wheel arches in the process, and then scrunched everything into a 170-in package. The whole thing was a mad assemblage of unrelated lines and shapes.

The styling could have been forgiven, or at least overlooked, had the 200-SX been the screamer that the concept—and Datsun ads—promised. The 200-SX was, according to the copy writers, "A sporty car with everything but a sports car price." Much ado was made about the engine. The 2.0-liter had a single overhead cam, "the type sports cars are made of." Yet the engine really came from a sedan, was rated only 1 hp more than the lamented 510 (though net, not gross), and even with the five-speed manual transmission could manage only a 19.5-sec quarter-mile and a 13.3-sec 0-60-mph dash. That—and the 200-SX's 2,365-lb curb weight—were simply a sign of the times.

Considering the humble origin of the 200-SX, its interior was remarkably upscale and, for the times, well designed. The instrument panel was an ellipsoidal recess across two-thirds of the dash. The driver was impressed with the

The 1977 200SX had the peculiar look of being designed by different departments that didn't talk to each other. *Automobile Quarterly Photo and Research Library*

car's full instrumentation, including a big 100-mph speedometer and 7,000-rpm tachometer, with oil pressure/fuel level and voltmeter/temperature gauges to either side. A cluster of warning lights included the "floor temp" light common when catalytic converters were first used. Datsun added to the mandatory cacophony of buzzers an annoying one of its own indicating reverse gear was engaged.

An advanced feature for the era, a Clarion AM/FM stereo radio was standard equipment, as were rear speakers. The 200-SX also had a center console with an electric clock, the manual shifter in a soft boot, hand brake lever, and armrest. There was also room for an optional cassette player. The three-spoked "sports" steering wheel had a padded vinyl rim, but the vinyl seats lacked side support.

The rear seat would generate protest from any adult who had to ride in it. Rear leg room was minimal and the roofline made rear head room even more limited. Even the front seats weren't overly kind to taller drivers. On the other hand, the front bucket seats reclined fully and cut-pile carpeting and tinted glass were standard. At least the trunk—small with a smaller opening—wouldn't encourage four-up touring.

The engine was fully serviceable and had a reputation for durability. Despite its 7,000-rpm capability, however, noise and vibration without added power encourage shifting at the 5,600-rpm power peak. The five-speed transmission was odd in that reverse was to the left and up, first to the left and back, as on race cars and Ferraris but inconvenient for day-to-day use. And shift feel garnered little praise, except that it showed off the engine's torque because it wasn't fun to shift. At least a three-speed automatic was optional.

Handling was disappointing, though perhaps not surprising. While well-sized steel-belted Toyo tires were standard and the recirculating ball steering light, the big engine, short wheelbase, and elementary MacPherson strut/live axle on leaf spring suspension conspired to create overwhelming understeer. Cornering power, at 0.658 g, was low even by '70s standards. And the wheelbase with short suspension travel meant ride was anything but limolike. At least the big disc/drum brakes were powerful and fade free even if somewhat numb.

Overall the 200-SX fell short of true Grand Touring status, something excellent construction and finish couldn't surmount. The base price of $4,399 compensated in part, about $1,000 less than a Toyota Celica or Volkswagen Scirocco, which were worth at least that much more. *Road & Track* summed it up saying, "For enthusiastic drivers the 200-SX just doesn't have the gearbox or handling to get the job done, so the word sporty must be confined to its visual image. For those who like its distinctive styling and aren't looking for over-the-road precision, it is an acceptable car at a good price."

With that faint praise, thank goodness the Datsun F-10 was uglier and that Nissan would in time keep the concept and fix its mistakes.

1978

The 200-SX was mostly unchanged, though the base price rose to $4,988. Emissions control took a further toll on the engine's drivability, if not actual rated power. The 1978 version of the engine, tested by *Road & Track* in a 1978 510 (which used, by the way, the five-speed manual with the "racing" shift pattern), "really feels its desmogging, backfiring on deceleration, not caring to rev over 5,000 rpm, lacking torque down low and seeming to run out of breath when strained." The late Seventies were bleak times for car enthusiasts.

1979-1980

The big change for the 200-SX for 1979 was a new black rocker panel, though horsepower fell to 92 bhp at 5,600 rpm and torque to 107 ft-lb at 3,200 rpm. The compression ratio had remained at 8.5:1, but tightening emissions controls took their toll in the twilight of mechanical (points and carburetor) engine management. Inflation in the Carter administration pushed the 200SX's base price to $6,229, an increase of more than 40 percent from two years earlier.

In its 3 1/2-year market run, the first generation 200-SX sold nearly 150,00 units, and almost two-thirds of them in the United States. So while it may have been a dud with the critics, it was not wholly rejected by the carbuying public. You can write your own moral to this story.

What to Look For

When *Car and Driver* first wrote up the new 1980 Datsun 200-SX, little care was taken to spare any feelings about its predecessor: "Cute grotesquerie is out, clean lines are in. The old car had all the outward visibility of a mine shaft and a dash whose lopsided plastic form must have passed near an open inferno somewhere along the production line. As a specialty car, it had little to recommend it unless you had a soft spot for sheetmetal freaks."

Little has changed since then. The car still looks peculiar and it hasn't become faster or more agile in the meantime, either. While a collector with a sense of humor, some overwhelming sense of loyalty, or a truly profound dedication to history—*some*one has to save one—might want a first generation 200-SX in the inventory, don't look for an Edsel effect to make the model more popular or more valuable. Interest in the first-generation 200-SX is so low that scarcity won't ever help the value or resale prices. On the other hand, if you want to be sure to have a one-of-a-kind at a car show, the first 200-SX is a sure winner.

For those who can't resist, the L20B has the same reputation for ruggedness as the rest of the Nissan L-series engines. And the suspension and drivetrain are straightforward and generally available, thanks to interchangeability with other Nissan products, over the counter from Nissan or at your local auto parts store. Watch out for that peculiar five-speed transmission.

1977 Datsun 200-SX
Base Price: $4,399

General

Layout	rwd
Curb weight (lb)	2,365
Wheelbase (in)	92.0
Track, front/rear (in)	50.4/49.8
Overall length (in)	170.0
Width (in)	63.0
Height (in)	51.2
Ground clearance (in)	n.a.
Seats	4
Trunk space (cu-ft)	n.a.
Fuel tank capacity (gal)	15.8

Chassis and Suspension

Frame type	Unit steel
Front suspension	MacPherson struts, lower lateral links, compliance struts, coil springs, anti-roll bar
Rear suspension	Live axle on leaf springs, tube shocks
Steering type	Recirculating ball
Steering ratio	n.a.
Tire size	175/70HR-13
Wheel size (in)/type	13x4 1/2/steel disc

Brakes

Front type	Disc
Size (in)	9.6 diameter
Rear type	Drum
Size (in)	9.0x1.58

Engine

Type	OHC Inline 4
Bore & stroke (mm)	85.1 x 86.1
Displacement (cc)	1,952
Compression ratio	8.5:1
BHP @ rpm	97 @ 5,600
Torque @ rpm (ft-lb)	102 @ 3,200
Fuel supply	Hitachi 2v
Fuel required	Regular

Drivetrain

Transmission speeds/types	5/man and 3/auto
Ratios 5-speed (:1)	
1st	3.38
2nd	2.01
3rd	1.31
4th	1.00
5th	0.85
Final drive ratio	3.89

Performance (Source: Road & Track, 5/77)

0-60 mph (sec)	13.3
1/4-mile (sec @ mph)	19.5
Top speed (mph)	98
Fuel economy (mpg)	23.5 observed

1980-1984 Datsun 200-SX

There aren't much bigger changes than this. Not much more than the engine block remained the same between the 1979 200-SX and the 1980 model. Not only was the styling changed, but so was the theme, from awkward curves to neatly creased straight lines. The old 210-based chassis was ditched in favor of one based on the new 510, the engine received more power and a cleaner exhaust (*sayanora* to the carburetor), and the interior was given a new motif—disco lite, perhaps, instead of restrained rocket boy.

It was nevertheless leagues ahead of the old 200-SX (at which the magazine reports on the new car couldn't resist taking a few final shots). The styling was the most obvious change but the chassis was the most important change in transforming the 200-SX. Not only was the wheelbase a precious 2 1/2 inches longer, but the MacPherson strut front suspension had an offset spring to counteract the natural bending forces on the strut. The springs were not angled however; the spring axis was set parallel to the strut rod. The old 210 leaf springs were gone in favor of a four-link live axle plus anti-roll bar arrangement.

The engine was another legacy of the new 510. The new 510 was, per *Road & Track*, possibly the "most average car ever built," an assessment that included its 2.0-liter, four-cylinder engine. In carrying it over to the new 200-SX, however, Nissan left little more than the crankshaft, flywheel, and oil and water pumps unchanged. Even the engine block was modified.

The 1980 200SX may still have been guilty of garish ornamentation, but it was improvement over its predecessor. A 1981 2-door hardtop SL is shown. *Nissan*

The 200SX was also available as a fastback, as seen by this 1981 SL example. *Nissan*

The 200SX hardtop coupe was revised in 1982. Note the grille, wheels, and B-pillar. *Nissan*

The aluminum cylinder head, however, was all new and was the integral part of NAPS-Z, for Nissan Anti-Pollution System.

The system, which mostly consisted of the head redesign, involved a major shift from L-series tradition. Instead of having intake and exhaust on the same side of the engine, the new single overhead cam (SOHC) head was a crossflow design, with valves opened by rocker arms crossing under the camshaft from either side of the head. The combustion chamber itself was "semi-spherical," valves offset instead of opposed as in a true hemi-head. The advantages it drew from the hemi design were improved volumetric efficiency from the bigger possible valves with less shrouding and a shorter flame path from the spark plug.

NAPS-Z used a lot of recirculated exhaust gas to reduce NO_X emissions by keeping combustion temperatures lower during part-throttle operation. As much as 17 percent of exhaust gases were recirculated in California emissions-equipped cars, causing, in fact, drivability problems for the West Coast cars. So while 49-state emissions-equipped cars have a swirl-inducing blade at the entrance to the intake port, Nissan went to a dual-plug head for the California cars, adding another spark plug to each cylinder just opposite each valve. Gone were the expected problems of misfiring, bogging, poor fuel economy, and difficult cold running. In fact, federal and California cars had the same power and torque ratings, both improved from the prior year's, at 100 bhp at 5,200 rpm and 112 ft-lb of torque at 3,200 rpm. Fuel injection, similar to that used on the 280ZX, was more than partly responsible for the power increase. For comparison, the 510 got NAPS-Z but without fuel injection and made eight fewer horsepower.

The old gearbox, with its "racing style" five-speed shift pattern, was replaced by a new five-speed, essentially Z-car gearing in a redesigned case. Overdrive fifth gear was now to the right and up, although a rather long reach, and reverse was straight back from that with no lock-out mechanism. Another new treat was four-wheel disc brakes instead of the old disc/drum format. The 1980 200-SX came with Bridgestone or Toyo 185/70SR-14 radials on 14x5-in steel wheels with full wheel covers.

The new 200-SX's stylists, however, seemed to have come from another planet, one closer to Earth, than those who did the old car. Whether notchback coupe or fastback, the new design was trim and crisp, definitely Guigiaro influenced. The federal bumpers finally looked as if they had been designed with, rather than after, the rest of the car, which had a tidy, integrated look. In the style of the day, corners were squared—even quad rectangular headlamps are used—and tricks were done with the rear side windows. The notchback carried a large window in the C-pillar, good for outward visibility but still are as dated as a *Charlie's Angels* rerun. The fastback had slanted louvers

The 1982 hatchback received detail changes. Note the B-pillar, wheels, and taillights of SL model shown. *Nissan*

back there and would have looked great with your Farah Fawcett 'do. Nissan also offered two-tone paint schemes, which proved to be love-it-or-hate-it.

The interior showed the disco era influence with velour upholstery (changed to a more professorial corduroy by the end of model run), though color choices and coordination in interior plastics were brought to a new high. Contrary to what enthusiast drivers would want, Nissan's interior designers went for a steering wheel with spokes at five and seven o'clock. The front seats were bolstered for good lateral support and had multiple adjustability, and the rear seats had more room than before but were still not the place for transporting adults long distances. The seat bottoms were low to the floor and headroom, especially in the fastback, was limited. For those with things to haul, the hatchback's rear seatback was split for either side to fold forward.

The dash continued the angular themes of the exterior. The instrument panel lived under a rectangular hood and in addition to full analog dials there was a "Multi-Warning System" (on SL models) for low fluids in the fuel tank, battery, radiator, and windshield washer reservoir. Power mirrors were standard and the SL's Clarion AM/FM stereo radio had a joystick to balance the fourspeaker system.

The base (Deluxe) 200-SX was fairly well-equipped, but the $800 SL package added the AM/FM stereo with power antenna and four speakers, a rear wiper/washer, and raised-white-letter tires. A sunroof was a $190 option, air conditioning added $580, and the three-speed automatic went for $260. Overall, the notchback listed for $6,289, the hatchback for $200 more.

Road tests generally lauded the 200-SX as much improved over its predecessor and approved its specifications. But the engine, satisfactory around town, became harsh and noisy between 4,000 rpm and its 6,300 rpm redline, and the car wasn't particularly fast. The suspension also worked well at slow speeds and in high-speed straight line driving, but the 200-SX understeered with a mania until the inside rear wheel lifted off the pavement and the rear end threatened to come around. Stopping distances were about average from the all-disc system, which demonstrated a surprising tendency to fade.

The overall result was a car that, while pleasant enough and attractive, disappointed enthusiasts with behavior somewhat short of sporting. Its *sporty* looks and ambiance, however, probably were enough to ensure its success. Nissan, after all, defined the 200-SX not as a sports coupe or even sporty coupe, but as a

Note the new wheels on 1983 200SX hatchback. *Nissan*

personal/luxury automobile. And at that it was very close to the mark.

Base price for the 1980 200SX coupe was $6,389, the hatchback selling for $200 more. The SL package added $800 to either. The automatic transmission added $250. Nissan sold 92,514 of the first-year 200-SX.

1981

There were no major changes for the second-generation's second year, although the twin-plug NAPS-Z engine was made standard for all 200-SXs, not just those sold in California. Features made standard on the SL for 1981 included power steering, power windows, power antenna, an upgraded radio, and a quartz digital clock with date and elapsed time functions. Styling changes included a new steering wheel and longer door armrests with power window and mirror controls. The SL models had new headliner and seat cloth, and color coordination and footwell lighting were improved. Halogen sealed-beam high-beam headlamps were also included in the SL package. Identifying an SL was made easier by bright trim on the headlight surrounds, tail pipe extensions, and on the hardtop's taillight bezels.

Optional on the hardtop were "luxury" alloy wheels and white sidewall radial-ply tires; the fastback could have "sporty" alloy wheels with raised-white-letter tires. Nissan also expanded the availability of the popular two-tone paint schemes, adding three more combinations.

Base price for the coupe was $7,389, and the hatchback cost $200 more. Sales for 1981 reached 76,024 units.

Sumptuous leather upholstery was optional on the 1983 200SX. *Nissan*

1982

In 1982 the 200-SX received the third-year spiff up that car makers use to make buyers of first-year models realize their cars are no longer new. It had to look different, so Nissan gave the '82 200-SX color-integrated "soft" urethane bumpers, an egg-crate grille, reshaped hood, new taillamps, and horizontal rather than vertical louvers on the B-pillar.

More power wouldn't hurt, so the 2.0-liter four had a new bore and stroke that resulted in 2.2 liters. Horsepower went up only by two, to 102 bhp at 5,200 rpm, but torque received a useful increase to 129 ft-lb at 2,800 from 112 ft-lb at 3,200 rpm, making the torque curve broader and taller. That dropped the quarter-mile e.t. by about a second, but the engine still wasn't particularly fun to run, again getting harsh and noisy above 4,000 rpm.

Sporty driving wasn't very rewarding either. The car still understeered and embarrassingly lifted its inside rear wheel in a corner.

Nissan added the vocal warning feature for 1982, a digital recording of a female voice instead of warning buzzers. You've heard the jokes: "Your door is ajar." There was no warning voice to say, "This isn't a sports car." Personal/luxury reigned again in 1982.

Personal/luxury wasn't cheap. The coupe had a base price of $7,939, with the hatchback again $200 more. The SL package added from $1,350 to $1,570 more. Sales totaled 48,559 units.

1983

Power steering and a stereo radio were features to become standard on all 200-SX models in 1983. Price edged up only slightly, only $200 more than the prior year. Sales, however, dipped to 31,158.

1984

The 1984 200-SX was a carry-over from 1983, destined to be dropped in mid-year with the introduction of a third-generation 200-SX. For the final go-around, prices climbed to $8,499 for the coupe and the usual $200 more for the fastback. The sales total for new and old 200-SX for 1984 was 63,466.

What to Look For

Although much improved over the 1977-1979 200SX, don't expect the 1980-1984 models to be more collectible. The first generation at least has funk on its side, but the second generation was relatively bland by comparison. So while the first 200SX may develop some cult status and be fun to take to car shows, the second generation will just be an older car for much longer. On the other hand, the later 200-SX will make a much better used car, better riding and with the NAPS-Z engine, better running.

1980 Datsun 200SX SL hatchback
Base Price: $6,489

General

Layout	rwd
Curb weight (lb)	2,690
Wheelbase (in)	94.5
Track, front/rear (in)	53.0/53.0
Overall length (in)	176.4
Width (in)	66.1
Height (in)	51.6
Ground clearance (in)	6.7
Seats	4
Trunk space (cu-ft)	12.4 10.8
Fuel tank capacity (gal)	15.9

Chassis and Suspension

Frame type	Unit steel
Front suspension	MacPherson struts, lower lateral arms, compliance struts, coil springs, anti-roll bar
Rear suspension	Live axle, lower trailing arms, upper angled arms, coil springs, tube shocks, anti-roll bar
Steering type	Recirculating ball
Steering ratio	18.0-20.5
Tire size	185/70SR-14
Wheel size (in)/type	14x5/steel disc

Brakes (vacuum assist)

Front type	Disc
Size (in)	9.9 diameter
Rear type	Disc
Size (in)	10.6

Engine

Type	SOHC Inline 4
Bore & stroke (mm)	85.1x86.1
Displacement (cc)	1,952
Compression ratio	8.5:1
BHP @ rpm	100 @ 5,200
Torque @ rpm (ft-lb)	112 @ 3,200
Fuel supply	Bosch L-Jetronic
Fuel required	Regular unleaded

Drivetrain

Transmission speeds/types	5/man and 4/auto
Ratios 5-speed (:1)	
1st	3.32
2nd	2.08
3rd	1.31
4th	1.00
5th	0.86
Final drive ratio	3.70

Performance (Source: Road & Track, 12/79)

0-60 mph (sec)	11.6
1/4-mile (sec @ mph)	18.8
Top speed (mph)	110
Fuel economy (mpg)	25.0 observed

Chapter 14

★	1984 1/2-1988 200SX
★1/2	1984 1/2-1986 200SX Turbo
★1/2	1987-1988 200SX SE

1984 1/2-1988 Nissan 200SX

The third-generation 200SX debuted at the Chicago Auto Show in February 1984, making it a half-year model. Both it and the second generation were called "1984 models," and if that now causes any confusion in the classifieds, there was no doubt then—or now when viewed in the sheet metal—over which is which. The 1984 200SX—the half-year model—was almost as much an advance over its predecessor as that car had been over the first 200SX.

First evidence lay with the new look. Corners were more rounded, but the wedge shape was still there, even more emphasized by a hood that dropped almost to the bumper.

The 1984 200SX Hatchback Turbo showed the model's new face—and a huge, slab-like hood bulge. *Nissan*

Styling of the 1984 200SX coupe was considerably subdued from its predecessor. *Nissan*

Pop-up headlamps were required to meet lighting height requirements (and followed the current fashion, e.g., Prelude and Celica). Just above the bumper, a narrow grille stretched from turn signal to turn signal but most cooling air entered under the bumper itself.

Fastback and notchback—"coupe"—versions were offered again, the styling of both much cleaner and simpler. The fastback lost its fussy B-pillar and three-side window arrangement in favor of a simple two windows per side. The coupe's treatment of the rear-quarter window was probably too ornate, though fitting for the market desiring a "formal" roofline.

The new bodies not only looked better but worked better, too. The new shapes plus flush-mounted glass, the retractable headlamps, semi-concealed wipers and airdams—and a decklid spoiler on the notchback—reduced the drag coefficient (Cd) of the fastback to 0.34 from the prior model's 0.47. The coupe's Cd took an incredible dive from 0.42 to 0.34. The car was also slightly smaller, with an inch-longer wheelbase. The new cars were 2 in shorter and 1/2 in narrower, but 0.8 in taller.

Naturally Nissan redid the interior. Both coupe and fastback were still 2+2, with the latter still meaning occasional seating. Front headroom actually gained an inch but in the back, passengers of normal stature would have to scrunch and ride with their knees in the air. For carrying a lot of cargo, the hatchback had split folding-rear seats. The front buckets had bolsters with big wings on the seatback for good lateral support. Designers were listening to the needs of sporty drivers.

Interior styling went with the times. Like the exterior, edges were softened but there were still a lot of straight lines. The steering wheel hub looked like some sort of *Star Wars* squadron emblem but at least the spokes had moved up to 3:15 and a quarter to nine. The shifter had an unusual squares-in-squares plastic boot and, mostly for styling, a row of warning lights were grafted on the dash right in front of the front seat passenger, stylish perhaps but not very visible to the driver. The main instrument panel was shrouded by an angular hood. A 125-mph speedometer and 8,000-rpm tachometer were supplemented by lesser gauges including, on the Turbo, a boost gauge.

Strides were made in chassis design, too. Nissan resisted the trend to front-drive sporty cars, an easy way out that would have allowed sharing components with the Sentra, the front-drive replacement for the 510. Nissan, without an inexpensive sports car stablemate for the more costly 300ZX, consciously planned the 200SX, and especially the 200SX Turbo, to fill that void.

Rear drive was retained for its inherently better high-speed handling qualities, and parts inventory could be shared with the upscale

The 200SX Hatchback was also available in non-turbocharged form. *Nissan*

Maxima and, to some extent, with the 300ZX. The live rear axle on four links and coil springs was a carry-over from prior years (and a Maxima wagon piece). The angled upper arms remained as they had been, but the lower parallel arms were fitted with liquid-filled bushings on the five-speed cars for better road shock absorption.

Independent rear suspension (IRS) was a new feature for the 200SX, though only on Turbo models. The semi-trailing arm set-up, similar to that used on the 300ZX, had coil springs, gas-filled shocks, and an anti-roll bar. With IRS, the rear track was widened to 56.1 in (compared to 53.7 in for standard models and 53.5 in for earlier cars).

Front suspension was the ubiquitous MacPherson struts, similar to that used on 1983 models except for added caster for improved directional stability and other changes to reduce rolling resistance. Track was stretched to 54.3 in (from 53.0 in). The really big news was that rack-and-pinion steering replaced the old recirculating ball system, not only greatly improving feel but reducing the car's turning circle by almost 2 ft.

Bigger tires went on the Turbo. Standard models had 185/70SR14 rubber, while the Turbo had 195/60HR15 tires on 15x6-in cast-alloy wheels. Standard models dropped the four-wheel disc brakes of earlier models (probably superfluous on less-sporting models) for a disc/drum layout; the front discs, previously solid, were ventilated on all models. The Turbo, meanwhile, had discs, solid at the rear, at all four wheels.

Ah, the Turbo. Nissan openly stated that the company wanted to sell more 200SXs to men (or at least, per press materials, "masculine buyers"), and figured more power was the way to do it. The standard engine for both the hatchback and coupe was the 1,974 cc overhead-cam unit, designated CA20E, taken from its transverse mounting in the Stanza and set longitudinally in the 200SX. This wasn't the old engine but a smaller, lighter block that was designed, as Nissan pointed out, with finite element analysis. Quieter and lighter, belt drive replaced chain drive for the overhead camshaft. Displacement was reduced, the naturally-aspirated engine's nominal 2 liters down from 2.2 liters the year before. Bore and stroke became decidedly undersquare at 84.6x88 mm to maintain proportions necessary for a fast-burn engine's compact combustion chamber. Maximum power stayed at the 102 bhp of the old 2.2 engine. The five-speed was standard, while a four-speed automatic with lock-up torque converter was a new option with the non-turbo engine.

The Turbo engine, called CA18ET, was the same as the naturally aspirated engine, though with a small assortment of changes to handle the turbocharger and the power it produced. Displacement was reduced to 1,809 cc, purportedly because the engineers wanted to keep fuel economy for the turbocharged engine close to that of the non-turbo engine. The compression ratio was dropped from 8.5:1 to 8.0:1 by increasing volume in the head and piston faces. For durability, new piston rings were used. The AiResearch T2 turbocharger had a water-cooled bearing to prevent coking of the lubricating oil. Boost was limited to 7.0 psi by the turbo's integral wastegate. The Turbo used Nissan's ECCS engine control system, a combination of a Hitachi hot-wire airflow sensor, a knock sensor, an exhaust-oxygen sensor, and water temperature sensor, and other monitors analyzed by a microprocessor to run the engine's fuel injection, spark timing, and idle speed. The Turbo made 120 bhp at 5,200 rpm and 134 ft-lb of torque at 3,200 rpm.

The turbo engine came only in the fastback and there would be little doubt from anyone that this was the hot rod. Nissan added a front air dam, rocker panel extensions, and rear deck spoiler, all in black, along with the window trim. "TURBO" was lettered on each end of the rear spoiler, presumably for the benefit of drivers being passed. Turbo models also were adorned with a hood bulge, which Nissan protested as necessary to provide clearance for the turbo motor, but was nevertheless criticized as a non-functional hood scoop disrupting the lines of the car.

The Turbo was naturally the favorite of enthusiasts. Performance of the naturally aspirated models was much like that of the earlier cars, but the Turbo could pop off quarter-miles in under 17 sec and go zero-to-sixty in less than ten. The 200SX didn't have the explosive feel of some turbos either, the engine having a strong surge all along the torque curve. Testers bemoaned the fact that the independent rear suspension increased weight by some 300 lb, in one regard making the turbo motor necessary to pull the extra weight around.

At least the suspension improvements

The hood bulge was shaved off the Turbo and its rear spoiler, shown here on a 1986 Turbo, with a centered, high-mounted stop light embedded in it, was extended to all hatchback models. *Nissan*

showed up not only in skid pad numbers but also in balance and feel. No more massive understeer, rather it had the front-to-rear equilibrium that a front engine/rear drive car was supposed to have. With four-wheel discs and lighter overall weight, braking performance was improved, though some testers reported the rear discs had too much tendency to lock when braking from high speeds.

Turbo models included a theft-deterrent system as standard equipment. Sunroofs were optional on all hatchback and notchback models, although the fastback's was the lift-off type while the notchback had a glass power-operated sunroof.

An XE trim level added a tinted upper windshield band, dual electric side mirrors, power windows, variable speed wipers, and cruise control. The XE group also added the "voice warning system," though for the first time including a switch to tell Miss Nissan to be quiet. Optional on XE models was a power pneumatic six-way adjustable driver's seat and a digital instrument cluster that was bundled with a trip computer, cassette player, sunroof, and alloy wheels. The digital dash was another example of why they became unpopular, the tachometer a horizontal bar graph of lights at the bottom of the instrument panel, not only hard to see but seeming to lag behind the engine revs as well. On notchback XE models a power woofer speaker system was standard unless the digital gauge package was installed.

MSRP for the notchback in base ("Deluxe") trim was $8,499, and the hatchback was the usual $200 more. The XE trim versions list for $9,699 and $9,999, respectively, while the Turbo topped the line at $11,699. Air conditioning added $640. Sales for all 1984 200-SX models totaled 63,466, about doubling the prior year's sales.

1985

There was little need to change the 200SX for 1985 after a mid-1984 introduction. Prices, though, continued their upward climb, the Deluxe notchback and hatchback up $500, the XE models up $550 and $750, respectively. The base Turbo hit $12,349. Second-year sales surely delighted Nissan executives, edging 1984's at 65,792.

1986

Nissan's motto for the 1986 200SX might well have been "leave well enough alone," with no engine or drivetrain changes. Nissan, however, did make the independent rear suspension and four-wheel disc brakes available on all 1986 200SX models.

An "Automatic Interval Wiper System" was available with the digital equipment package. The system had a hood-mounted sensor that measured the "frequency of moisture impact," with the interval windshield wiper control module then varying wiper frequency when the wipers were set in "interval" mode.

The interior got a new steering wheel and new fabrics and the Turbo's rear spoiler—extended to all hatchback models—was restyled with the center high-mounted stop light embedded in it. Nissan must have solved those mechanical clearance problems because the hood scoop was gone. Bigger tires—195/60R15 on non-turbo, 205/60HR15 on turbo models—followed a trend that was as inevitable as rising prices. The 1986 Turbo listed for $13,698.

The V-6 fastback SE model replaced the Turbo in 1987; all models received the reshaped front spoiler and rocker panels plus a body color (instead of black) rear spoiler. *Nissan*

1987-1988

Turbo came and turbo went. The optional performance engine for the 1987 200SX was a V-6. And not just a V-6, but a V-6 that was bigger again by two-thirds, with a third more horsepower. The 120-bhp turbo four had given way to a 160-bhp six, and despite the turbo's reputation for having a relatively non-explosive power curve, the six promised smoother power delivery and a less-frantic experience at the helm.

Nissan didn't invent a new engine for the 200SX, merely borrowed the naturally aspirated version of the V-6 used in the 300ZX since 1984. This 2,960-cc six had a cast-iron block and aluminum heads with one chain-driven overhead camshaft per head.

Like the earlier turbocharged engine, the V-6 was available only in the fastback. Though Nissan naturally revised its model designations, the "Turbo" title was no longer very appropriate. The V-6-powered models received an "SE" suffix, while the carried-over four-cylinder adopted "XE" (formerly a trim level) in either coupe (notchback) or fastback (hatchback) guise.

A new disguise also accompanied both SE and XE models in 1987, a simpler look with the front spoiler (reshaped), side sill extensions (ditto), and rear spoiler no longer in black but in the body color. Taillamps were reshaped and the SE's aluminum wheels were a new disc-with-perimeter-slots design, but wheel and tire size remained the same.

The interior got a once-over, with a new three-spoke steering wheel and new fabrics and changes in some of the soft trim, but gauges and controls stayed as they were.

The V-6 engine wasn't completely without drawbacks. The bigger engine and stronger drivetrain components made the SE some 200 lb heavier than the Turbo, edging curb weight up to 2,976 lb. A *Car and Driver* road test showed its test 200SX SE to only equal the Turbo in acceleration, which the magazine attributed in part to the extra poundage. On the other hand, with no changes in specifications, the 200SX SE stopped fast and true, Nissan apparently having fine-tuned the four-wheel disc brakes. Handling, though, was so-so, with "no nasty vices" but no outstanding merits either.

The 1988 model year brought new wheels for the SE, plus a leather-wrapped four-spoke steering wheel. Prices continued to climb, with the 1987 tags at $10,849 for the XE coupe, $11,199 for the XE hatchback, and $14,499 for the SE hatchback. Prices for 1988 jumped to $12,449 for the XE coupe, $12,749 for the XE hatchback, and $15,549 for the SE hatchback. Sales followed the normal decline for a model that has been on the market for several years: 1987 sales of the 200SX totaled 30,196 units. Combined sales for the 1988 200SX and 1989 240SX (calendar year 1988) amounted to 30,196 units.

What to Look For

The styling of the 1984-1988 Nissan 200-SX has weathered well, and even at ten or so years old doesn't look terribly obsolete.

Nissan engines are notoriously rugged, although the usual warnings apply as with any turbocharged engine. Poor maintenance will wreck it, and a turbo motor in poor condition will be expensive to repair, particularly for those who have to hire someone to disconnect and reconnect all those parts. Turbos are more likely to have been driven hard. Take care.

1984-1/2 Nissan Turbo XE fastback
Base Price: $11,699

General	
Layout	rwd
Curb weight (lb)	2,830
Wheelbase (in)	95.5
Track, front/rear (in)	54.3/53.5
Overall length (in)	174.4
Width (in)	65.4
Height (in)	52.4
Ground clearance (in)	5.3
Seats	4
Trunk space (cu-ft)	11.8 8.7
Fuel tank capacity (gal)	14.5

Chassis and Suspension	
Frame type	Unit steel
Front suspension	MacPherson struts, lower lateral links, compliance struts, coil springs, anti-roll bar
Rear suspension	Semi-trailing arms, coil springs, tube shocks, anti-roll bar
Steering type	Rack & pinion, power assist
Steering ratio	16.4:1
Tire size	195/60R-15
Wheel size (in)/type	15x6/cast alloy

Brakes (vacuum assist)	
Front type	Vented disc
Size (in)	9.8 diameter
Rear type	Disc
Size (in)	9.0 diameter

Engine	
Type	OHC turbo Inline 4
Bore & stroke (mm)	83.0 x 83.6
Displacement (cc)	1,809
Compression ratio	8.0:1
BHP @ rpm	120 @ 5,200
Torque @ rpm (ft-lb)	134 @ 3,200
Fuel supply	Bosch L-Jetronic
Fuel required	Regular unleaded

Drivetrain	
Transmission speeds/types	5/man and 4/auto
Gear ratios 5-speed (:1)	
1st	3.59
2nd	2.06
3rd	1.36
4th	1.00
5th	0.81
Final drive ratio	3.90

Performance (Source: Road & Track, 5/84)	
0-60 mph (sec)	9.5
1/4-mile (sec @ mph)	16.9 @ 80.5
Top speed (mph)	116
Fuel economy (mpg)	20.5 observed

Chapter 15

1989-1994 Nissan 240SX

Nissan confounded industry watchers. The successor to the 1984 1/2-1988 200SX should have had front-wheel drive. Some said the earlier '84-1/2 200SX—like almost all of its competitors—should have switched to the front. But the new 240SX, named after its engine displacement (although if it made one think "240Z," that was OK, too), stayed with the venerable *systeme Panhard*: front engine, rear drive.

The objective was balance. Front-wheel-drive sports coupes may be more economical in borrowing engine, drivetrain, and suspension

The new 1989 240SX was offered in coupe form, as shown here in XE trim. *Nissan*

bits from front-drive sedans—where the space efficiency of front-drive is valuable—but better handling came from apportioning steering front and drive rear. Nissan's product planning guru was quoted as stating, "That layout is consistent with real sports cars. We wanted to keep the balance and the handling. With a weight distribution of 52/48 percent, I know at least we have the balance."

What indeed had Nissan forged? It was, for starters, a new car from the ground up. Nissan kept MacPherson front suspension although it threw in more anti-dive geometry. At the rear, however, was a new multi-link suspension. Upper and lower diagonally located A-arms combined with an upper lateral link to dial in increasing toe-in as cornering loads grow, providing a form of passive rear steering. The Cray supercomputer-developed geometry also minimized squat, lift, camber change, and jacking without having to resort to stiffer springs or bushings.

Nissan figured it lost sales in '88 to high insurance rates, so it settled on one engine for United States-bound 240SXs, a decision that was made with the advice of American insurance companies who blanched at the thought of a V-6. The 2,389-cc 12-valve (two intake, one exhaust) single-overhead-cam (SOHC) inline four was rated at 140 bhp, a setback from earlier models. The new 240SX was capable of 8.8-sec 0-60 mph sprints in 240SX; the old V-6 had been marginally quicker at 8.6 sec. The 240SX also was governed to a 112-mph top speed, again to keep insurance costs down. In Japan a 1.8-liter, 133-hp 16-valve, twin-cam four was standard with an optional 172-hp version available.

Vibration in the big four was combated by finite element analysis of the engine. Nissan

A sleek 240SX fastback was new in 1989. An SE model shown here. *Nissan*

An optional instrument package offered in 1989 included a digital speedometer that had a head-up display projected on the windshield. *Nissan*

declined to used balance shafts, which add weight, and instead strengthened the crankshaft bearing webs to make the block more rigid, reducing vibration and noise. New noise-reducing technology let Nissan opt a for long-lived chain drive for the camshaft in lieu of rubber belts that require periodic replacement.

The all-new KA24E engine was developed with the goal of a broad torque band, the 4,400-rpm torque peak somewhat deceiving in that substantial torque was available at lower rpm. Maximum torque was still, at 152 ft-lb, 22 down from the earlier six. Redline was 6,400 rpm.

A new five-speed manual transmission, with double-cone synchromesh second gear, was standard, and a four-speed automatic was optional. The electronically controlled automatic had a lock-up torque converter to enhance fuel economy.

The look of the 240SX was as new as its underpinnings: longer, wider, and with a wider track and a wheelbase stretched by almost 2 in. The body lost its straight edges, however rounded, in favor of the organic look of the Nineties. The front end was sleek, with pop-up halogen headlamps and a small double opening above the integrated bumper. Two body styles, differing from the cowl on back, were offered. The Fastback, or SE for 1989, was the sportier of the two, with flush B-pillars and concealed C-pillars, while the notchback Coupe, or XE, offered "a sophisticated, fashionable alternative to those seeking the convenience of a separate trunk."

Nissan clearly envisioned two different roles for the two body styles, with different interior trim and option packages. The optional first-ever-in-a-mass-production-automobile head-up display dash—with centered analog tachometer and offset digital speedometer that could be projected onto the windshield—was available only on the XE notchback. On the other hand, only the SE fastback could be ordered with a sports package that included sports suspension, 6.0x15-in alloy wheels, 205/60HR15 tires (195/50R15 all-season M+S standard), front air-dam extension, rear air dam, and cruise control. For a change, the spoilers in the sports package actually did some good, lowing the Cd of the SE fastback from 0.33 to 0.31. The XE coupe had a Cd of 0.34.

The interior made great strides into modernity. The front bucket seats were "monoform," fabric-covered with no seams looking as if they had just been pulled from the mold. Coupe seats were upholstered in velour, fastback in jersey cloth, and the passive safety belts had an active lap belt with a passive "mouse-type" shoulder belt. The rear seat in both body styles folded for added luggage room but the 240SX still lived up to its 2+2 designation with limited rear seat room. The dash was highly sculpted with contours continuing into the doors, but the instrument panel lost a couple of minor gauges, with only temperature and fuel gauges left with the 8,000-rpm tachometer and 115-mph speedometer.

Journalistic response was equivocal, *Car and Driver* calling it "a car to lust over" but also noted that there were competitors with more horsepower, something—insurance companies notwithstanding—the 240SX needed. *AutoWeek* noted that "the 12-valve engine . . . isn't bad, but like all big fours is a bit coarse compared to the smaller Toyota or Honda 16-valves, and it runs out of wind at about 5,000 rpm (600 rpm above the torque peak, 600 below the power peak). It just doesn't complete the excellent package."

This 1992 240SX LE fastback displays the new nose cap that was introduced the year before. *Nissan*

Handling delighted the automotive press, however. Although one critic called the '89 240SX a "devoted understeerer," another said its cornering reminded him of the Porsche 944. Yet again, another attributed "exceptional dynamics" to the 240SX, providing "a terrain-hugging ride but masterful control . . . its deft controls and cheery bent for changing directions belie its mass, subtracting about 400 pounds from its feel."

Base price for the XE notchback was $12,999, for the SE fastback $13,199, rising, respectively, to $13,249 and $13,499 in mid-model year. The automatic transmission added $830, air conditioning another $825. The sport package for the SE was $799, the pop-up sunroof $450 more. Nissan expected to sell few bare-bones 240SXs, however, anticipating most to be optioned up to around $16,000.

Nissan had great expectations for sales volumes for the 240SX, anticipating some 60,000 out the door in the first model year. The actual count was above that, at 68,993. About half had automatic transmissions and half had air conditioning; two-thirds were hatchbacks, of which half had optional ABS brakes, a mid-year addition available only on hatchbacks.

1990

Little was changed for the second year of the 240SX, sales for the first year having rung the cash register bells like a carillon. Nissan's changes in the 240SX were few: a power antenna was made standard, and the sport package of spoilers and such was made optional on the notchback XE as well as the fastback SE. With the sport package, the coupe's Cd went to 0.32.

Prices edged up to $13,249 and $13,499 for the notchback and the hatchback. Sales took a second-year slump, adding up to 47,816 for the model year. Eighty-three percent of these were fastbacks, while 41 percent of buyers went for the automatic.

1991

The 240SX followed the time honored rule of "freshening" a model in its third year, a sleek new restyled nose accompanied by more power, "enhanced suspension," and the option of Super HICAS four-wheel steering on the SE Fastback model.

Introduced the prior year on the 300ZX Turbo, Super HICAS steered the rear wheels in response to input of vehicle speed, steering angle, and rate of turn data. Not functioning as a low-speed maneuverability aid, the electronically controlled system turned the rear wheels out to aid turn-in and then turned the wheels back in for better stability through the corner.

Super HICAS was part of a $500 handling package that also included a viscous-coupling, limited-slip rear differential, stiffer suspension, and bigger 205/60R-15 non-M+S radials. That was surprisingly inexpensive, though buyers first had to sign up for air conditioning and the $995 anti-lock brakes.

More power came in the form of a twin-cam 16-valve head that increased output to 155 bhp at 5,600 rpm (an increase of 15 bhp) and 160 ft-lb of torque at 4,400 rpm (up eight). Nissan also improved the engine mounts and beefed up the block to quell vibrations, and though "hints (of) thrashiness" were noted above 6,000, the freer-breathing engine had a higher redline at 6,900 rpm. The power whittled 0-60 mph to 8.1 sec, down from 8.6 sec.

To keep everyone alert, Nissan reworked model nomenclature. Base models were known simply as the Coupe and Fastback. The SE models (in both body styles) had new alloy wheels, a rear wiper (on the fastback), cruise control, and power doors, locks and mirrors; the SE Coupe also had the digital HUD system as standard equipment. This was topped off by a "high image" LE Fastback model, which included the SE's good stuff plus leather seating surfaces and air conditioning as standard equipment. Anti-lock braking was optional on all SE models.

The new models were easily identified by

The 240SX convertible was introduced in 1992. Stubs on the doors are for seat belt mounts. *Nissan*

the new front-end styling, which repositioned turning lamps and air intakes for a smoother overall appearance that some loved and others thought too bland.

All prices were over fourteen grand for 1991, with the base Coupe and Notchback starting at $14,095 and $14,350, respectively. The SE version elevated the cost of entry to $16,200 and $16,390, and the LE rose to $18,170. It wasn't hard, with a few options, to have a 240SX that sold for more than $20,000. It was a long way from the budget sports car of just a few years before.

But U.S. model year sales stabilized at 38,991, of which 42 percent were equipped with automatic transmissions. About half had anti-lock brakes, more than two-thirds came with the factory aluminum wheels. A quarter were made with four-wheel steering. The most popular body style—78 percent—was the hatchback.

Nissan 240SX was honored as one of *Automobile* magazine's first "Ten Best All-Stars" in 1991 (when the magazine started the feature), with Douglas Weisz stating, "Rear-wheel drive, a responsive high-revving engine, and All-Star good looks add up to one of the best answers available today to the sports-car question."

1992

With the substantial changes in 1991 (as well as the honors), it's not surprising that 1992 was largely a carry-over year, except that there was exciting mid-year news. Before the July introduction of the 300ZX convertible (as a 1993 model), Nissan brought out a topless 240SX in April. Like the Z, the 240SX Limited Edition (only 20,000 copies were planned) Convertible was developed with ASC in the United States. Unlike its bigger brother, the 240SX was finished by ASC in Carson, California, as Nissan had installed some of the extensive reinforcements in Japan and the rest was done before ASC put on the top.

Also unlike the 300ZX, the top was power-operated. After the release of two latches on the windshield header, a push of a button on the dash collapses the well behind the rear seats (which were retained and snugger than ever). The top takes about 30 sec to drop but longer than that to install the soft boot. Posts were added to the doors for the shoulder belts, putting them up where they would do some good, but the stubs interfered with the car's shoulder line and caused buffeting.

The Convertible came well equipped, an automatic being part of the package—reportedly because even with reinforcements the chassis would not stand the stresses a manual transmission would inflict. A rear deck spoiler, alloy wheels, power windows and door locks, a leather-wrapped steering wheel, and a four-speaker stereo were standard on the Convertible. Air conditioning and a Sony CD player were the only options. Alas, the 240 gained 321 lb in the conversion and the mandatory automatic didn't help performance either, making the Convertible lose the crispness of its hard-topped counterparts. It wasn't the most rigid of platforms either.

The top looked handsome when raised, and had a neat and cushy headliner that took the Convertible out of the "ragtop" class. Unfortunately the rear window was plastic and subject to deterioration even if coddled. The convertibles also had a tendency to leak, enough so that Nissan issued a Technical Service Bulletin about a fix.

Even at a $21,995 base price, Nissan sold 2,102 of the 1992 model year 240SX Convertibles, about seven percent of the 27,300 1992 240SXs sold. Coupes (with a $14,515 base price, $16,690 for the SE) sold 8,518 units. Some 16,680 hatchbacks were sold, with $15,265 as the base price, SEs selling for $17,435, and LEs at $19,320.

1993

With the convertible introduced in mid-1992, there were no obvious alterations to the 240SX for the 1993 model year other than the addition of new exterior colors and some changes to option availability: Leather seats, previously the major part of the LE trim group, became a free-standing option, and anti-lock braking, previously bundled with the handling group, was also available individually. A technical change in 1993, beginning with VIN JN1MS34PXPW313771 and JN1M36PXPW313771, was a new cylinder head, marked by elimination of the Swirl Control Valve. New valve clearance specifications were issued for the new head.

Prices went up, but only slightly, with the notchback at $14,895 and $17,490, for base and SE models, respectively. Hatchbacks sold for $15,715 and $17,950, while the Convertible came in at $22,690. For the 1993 model year, sales of the 240SX totaled 23,587. Thirty percent were notchbacks, 53 percent were fastbacks, and 16.5 percent (3,892 units) were convertibles. Transmission installation was about 50-50, with slightly more five-speed manuals sold. Nissan changed the 240SX to non-CFC refrigerant in the air conditioner in mid-year.

1994

At the end of the line for the 240SX, Nissan dropped both coupe and fastback models for 1994, keeping only the convertible and pricing it at $23,969. The new 1995 240SX debuted at the 1994 New York Auto Show in April, going on sale shortly thereafter. It was time for a change.

What to Look For

The 240SX was reliable although there were some problems noted by Nissan in Technical Service Bulletins. The automatic transmission could slip in reverse or produce excessive shift shock, while the manual could develop 5th-gear vibration between 45 and 65 mph. The clutch could also judder and have excess wear. The rear final drive oil seal could also develop leaks. An engine oil leak near the oil filter and pressure switch has also been noted. Watch also for excessive wear in the poly-groove engine accessory drive belt.

As noted above, the convertible top can develop leaks. Look for water damage in the interior that can develop if the leak is not corrected. Look particularly in the headliner and lamp.

1989 Nissan 240SX SE hatchback
Base Price: $13,199

General	
Layout	rwd
Curb weight (lb)	2,800
Wheelbase (in)	97.4
Track, front/rear (in)	57.5/57.5
Overall length (in)	178.0
Width (in)	66.5
Height (in)	50.8
Ground clearance (in)	5.7
Seats	4
Trunk space (cu-ft)	12.8 8.0
Fuel tank capacity (gal)	15.9
Chassis and Suspension	
Frame type	Unit steel
Front suspension	MacPherson struts, lower A-arms, coil springs, anti-roll bar
Rear suspension	One diagonal link, two lateral links, and one control arm per side, coil springs, anti-roll bar
Steering type	Rack & pinion, power assist
Steering ratio	17.1:1 std./14.9 opt. Super Hicas
Tire size	195/60R-15
Wheel size (in)/type	15x6.0/cast alloy
Brakes (Vacuum assist; ABS optional)	
Front type	Vented disc
Size (in)	9.9
Rear type	Disc
Size (in)	10.2
Engine	
Type	SOHC 12-valve Inline 4
Bore & stroke (mm)	89.0 x 96.0
Displacement (cc)	2,389
Compression ratio	8.6:1
BHP @ rpm	140 @ 5,600
Torque @ rpm (ft-lb)	152 @ 4,400
Fuel supply	Nissan port EFI
Fuel required	Regular unleaded
Drivetrain	
Transmission speeds/types	5/man and 4/auto
Ratios 5-speed (:1)	
1st	3.32
2nd	1.90
3rd	1.31
4th	1.00
5th	0.76
Final drive ratio	4.11
Performance (Source: Road & Track, 1/90)	
0-60 mph (sec)	8.8
1/4-mile (sec @ mph)	16.5 @ 84.5
Top speed (mph)	131
Fuel economy (mpg)	23.8 observed

1995-1996 Nissan 240SX

"Little risk of igniting passions"—as Kevin Smith described the new 1995 Nissan 240SX in *Car and Driver*—may work for econoboxes and family sedans, but what about the replacement for a model the company had unabashedly called a sports car? Isn't passion what a sports car is supposed be all about?

Maybe, but Nissan wasn't calling the new 240SX a sports car. It was a "classy sports coupe," and Nissan pegged 240SX buyers as "sedan avoiders": empty-nest baby boomers who want to look youthful, but apparently not necessarily act that way.

Nissan gave the 240SX its U.S. debut at the New York Auto Show in early April 1994, with nationwide sales beginning later that month. It wasn't as new as its predecessor had been in 1989. The engine was the same 155 bhp 2.4-liter four that had powered its namesake the year before, although with a more rigid cylinder head and liquid-filled engine mounts to reduce noise. Higher compression (from 8.6:1 to 9.5:1) plus a recurved higher-lift camshaft boosted low-range torque while leaving peak power at 155 bhp at 5,600 rpm and torque at 160 ft-lb at 4,400 rpm. The cam profiles were also changed

The 1995 240SX, shown here in SE trim, was offered only as a coupe. *Nissan*

In addition to badging, the SE can be identified by its 16-in alloy wheels and the rear deck spoiler. *Nissan*

to minimize noise between the lobes and the lifters. To further reduce engine noise, the intake duct was recontoured and a second resonator was added. Improved control logic in the engine control computer and a pre-catalyzer/dual oxygen sensor system as well as other refinements resulted in a decrease in emissions by 40 percent over its predecessor.

The chassis was redone, the new body having 50 percent more torsional rigidity and 100 percent greater bending rigidity despite a weight loss of about 80 lb, thanks to the use of Nissan's Cray supercomputer to optimize the structure. Side impact door beams meet 1997 federal side-impact regulations, and the body side outer panel is a single piece for better looks and corrosion resistance. Sandwich steel—fusible-insul (a type of asphalt) between two stamped sheets of steel—was used for the firewall to limit engine noise reaching the cockpit.

Wheelbase, front and rear track, and wheel travel were all increased, but bushings, springs, and shocks were all softer and the choice of tires made more for a soft and quiet ride than athleticism. Super HICAS rear-wheel steering was out, even on the "sporty" SE version that got a rear anti-roll bar, slightly stiffer shocks and 16-in alloy tires and wheels (15-in steel wheels were standard on the base model), but testers were disappointed in the feel and precision of the suspension.

Anti-lock brakes, paired with a limited-slip differential, was a stand-alone option.

Nissan, in describing the styling, used the word "elegant." Doubtless, the lines were intended to attract a more mature—older—clientele than its more aggressively profiled predecessor. More conservative, it had grille between its aero-styled (rather than pop-up) headlamps and a more-defined front bumper.

The larger exterior dimensions made more interior room possible, but if the rear seat was a little less confining for adults it was still no limo pretender. New seats with conventional upholstery replaced the cloth-bonded-to-foam buckets of before. The rear seatback folded in one piece with a trunk pass-through. The designers worked hard on the interior, creating a highly sculpted dash and console. Driver and passenger side airbags were standard. Drawing praise were the 8,000-rpm tachometer overlapped by a 120-mph speedometer, flanked by fuel and temperature gauges, and accessibility and operation of controls.

While the previous 240SX was available in coupe or fastback version—the latter preferred by 70 percent of U.S. buyers—the 1995 240SX would be offered in the notchback coupe format only. It seems Japanese buyers preferred

The instrument panel on the 240SX featured white-faced dials. *Nissan*

the notchback, so only that would be sent to America as well.

Nissan retained the base and SE trim designations, though the former was hardly a stripped-down version. Standard on every 240SX was a 50-watt, four-speaker, AM/FM/cassette stereo audio system, power windows, and dual power mirrors; a convenience

The 1996 240SX had a revised grille. Wheel covers identify this as the base model. *Nissan*

package bundled extra options for the base model. The SE model included a new remote locking system and a six-speaker, 160-watt audio system; this model was identifiable by its fog lamps and front chin and rear deck spoilers. The instrument panel on the SE had black-on-white gauges that went to white-on-black after dark. Other standard SE features included cruise control and power door locks, while leather seating was optional. The base price for the 240SX was $17,499; the SE had a base of $21,219.

Lighter and with a wider torque curve, the 1995 240SX was faster than its predecessors, yet reviewers found the personality transplant less than desirable, at least to the enthusiast. As Kevin Smith said in *Car and Driver*, Nissan "turned a car we found easy to like into one that's hard to love."

1996

A surprise for 1996 was a new grille design for the 240SX, surprising because the model was only one year old—or, a year and a half if you count the early introduction. Though easy enough to do, changing so soon was an aggressive marketing ploy for a relatively new model. In addition to the grille, which was swapped for a single mail-slot design, Nissan also changed to a "sportier" seat cloth and shuffled option packaging.

What to Look For

The 240SX has been relatively trouble free, though some problems have occurred with the new OBD II system, which is covered under warranty. Engine overheating problems have been addressed by a technical service bulletin to dealers.

1995 Nissan 240SX SE
Base Price: $21,219

General

Layout	rwd
Curb weight (lb)	2,860
Wheelbase (in)	99.4
Track, front/rear (in)	58.3/57.9
Overall length (in)	177.2
Width (in)	68.1
Height (in)	50.8
Ground clearance (in)	5.1
Seats	4
Trunk space (cu-ft)	9
Fuel tank capacity (gal)	17.2

Chassis and Suspension

Frame type	Unit steel
Front suspension	MacPherson struts, lower A-arms, coil springs, anti-roll bar
Rear suspension	Lower angled A-arms, lateral links, upper semi-trailing links, upper lat eral links, coil springs, tube shocks, anti-roll bar
Steering type	Rack & pinion, power assist
Steering ratio	n.a.
Tire size	P205/55VR-16
Wheel size (in)/type	15x6.5/cast aluminum

Brakes

Front type	Vented disc
Size (in)	9.9 diameter
Rear type	Disc
Size (in)	10.2 diameter

Engine

Type	DOHC 16-valve Inline 4
Bore & stroke (mm)	89.0 x 96.0
Displacement (cc)	2,389
Compression ratio	9.5:1
BHP @ rpm	155 @ 5,600
Torque @ rpm (ft-lb)	160 @ 4,400
Fuel supply	Nissan port EFI
Fuel required	Regular unleaded

Drivetrain

Transmission speeds/types	5/man and 4/auto
Ratios 5-speed (:1)	
1st	3.32
2nd	1.90
3rd	1.31
4th	1.00
5th	.076
Final drive ratio	4.08

Performance (Source: Car and Driver, 5/94)

0-60 mph (sec)	7.5
1/4-mile (sec @ mph)	15.9 @ 86
Top speed (mph)	118 (governor limited)
Fuel economy (mpg)	22 EPA city

Chapter 17

★	1983-1986 Pulsar NX
★1/2	1983 Pulsar NX Turbo

1983-1986 Nissan Pulsar NX

When the Nissan Pulsar NX wedged itself into the 1993 U.S. model lineup, it certainly wouldn't be confused with anything else. Its exterior was styled by the Ogikubo Design Center of Prince Motors Ltd., Nissan's wholly owned subsidiary, and it looked as if the clay model could have been done with a samurai sword (John Belushi in "Samurai Car Stylist"?). More flat planes on a single automobile would be hard to imagine. Sporty, yes, but it was love-it-or-hate-it from the start.

It wasn't the first Nissan Pulsar. The Datsun 310 series, sold here from 1979, was known in Japan as the Pulsar from the beginning. Nissan had used numbers instead of names in the '70s for the U.S. market, but as with the Sentra, Stanza, and Maxima, Nissan granted this sporty origami on wheels a name.

The wedgy Pulsar NX debuted in 1983. *Nissan*

Pop-up headlamps were the rage in the early Eighties, and allowed the NX to have a rakish nose—at least when they were down. *Nissan*

The Pulsar really didn't replace anything in Datsun's lineup. The Datsun 210 had been superseded by the Nissan Sentra, the (second-generation) 510 by the Nissan Stanza. But first 310 was a different sort of car than the new Pulsar, more of an econobox with a dash of sportiness. The new Pulsar, in comparison, had the sportiness ladled on.

Nissan touted the long—relatively speaking—hood, pop-up headlamps, and short rear deck, as well as flush drip rails, the sharply slanted (32 deg) windshield, the integrated bumpers, and front and rear spoilers (the latter a lip on the rear deck) as being visually "striking"—their word. But it also contributed to the Pulsar's 0.37 coefficient of drag, then fairly respectable. Nissan then added tape stripes to the hood, sides, C-pillar, and trunk, finishing off with full wheel covers that looked like main-frame computer tape discs.

The Pulsar's dash was simple almost to the point of stark, though it did have an 8,000-rpm tachometer. *Nissan*

Styling aside, the Pulsar wasn't a clean-sheet design. The chassis was based on the major floorpan stampings of the Sentra and used the sedan's front-wheel-drive and suspension pieces. At the front were lower A-arms and an anti-roll bar and semi-trailing arms; coil springs and tube shocks tagged along behind. Rack-and-pinion steering was power assisted and gave moderately good road feel and control. Pushed hard, on the other hand, the tires and suspension quickly became overworked and the car uncomfortable. *Road & Track* called it "a good five/tenths car."

Nissan used an enlarged version of the Sentra's E-series engine, an overhead-cam four with a crossflow head, semi-hemispherical combustion chambers, and a Hitachi dual-throat carburetor. The Sentra 1.5-liter was given a 6-mm-longer stroke to produce the 1.6-liter Pulsar engine, very oversquare at 76.0x88.0 mm. Compression was raised from 9.0:1 to 9.4:1, and all the changes together were good for 2 bhp, the Pulsar's engine good for 69 bhp. The standard transmission was a five-speed manual with a lock-up torque converter; a three-speed automatic transmission was optional.

The Pulsar was officially a five-seater, but it was a 2+2 at best. Six-footers, thanks to a roofline actually higher at the back than the front, had adequate headroom in the back seat, but there was little room for their legs. Perhaps Nissan saved the name Pulsar, which refers to a type of collapsed star, for a car for which it was most appropriate. At least the rear seatback folded forward to add 10.5 cu-ft of cargo capacity to the trunk's already copious 14.0 cu-ft.

The front seats were comfortable with more side bolstering than the car's meager cornering capability deserved. Vivid upholstery created as many debates as the exterior design, but at least the instruments and controls were straightforward and easy to use. Power steering, a lift-off sunroof, and AM/FM stereo cas-

The short-lived Pulsar NX Turbo was identified by "turbo" graphics. *Nissan*

sette unit were all included in the standard "Deluxe" trim level. Only three options were available: air conditioning ($610), alloy wheels ($385), and a three-speed automatic transmission ($350). Fourteen exterior colors were offered.

Base price for the Pulsar NX was $7,399, not econocar cheap but not enough to price the car out of reach of a predominantly single, young, and female market for which it almost seemed designed. The NX was joined mid-year by two- and four-door hatchback Pulsars, but the NX outsold both together 60 percent to 40 percent, and the hatchbacks—competing somewhat with other Nissan models—were dropped after a single model year.

Pulsar NX Turbo

The Pulsar NX quickly earned the reputation of, to use Nissan's innocent phrase, "a secretary's car." It's no wonder. With ugly-cute looks, a rather tame 69 hp and weak-kneed suspension, it was hardly going to tempt the average red-blooded American male out of his Monte Carlo—or three-series BMW, for that matter.

The solution, of course, was more power, and this being the early Eighties, the method of choice was turbocharging. Nissan used the 1.5-liter version of the Stanza's engine to affix an AiResearch T2 turbocharger, there reportedly not being enough metal on the 1.6 to stand the stresses of the turbo. The 1.5 itself needed a lot of work to accommodate the extra pressure. Pistons, rings, and wrist pins were stronger, along with bigger connecting rods and a modified crankshaft. The compression ratio was dialed back from 9.2:1 to 7.8:1 to compensate for the 6.8-psi maximum boost. A special exhaust manifold with individual runners separated exhaust pulses for better low-rpm boost, and exhaust valves were treated to withstand higher gas temperatures.

The turbocharger made things a lot hotter under the hood, so Nissan added a block-mounted oil-water heat exchanger and an auxiliary fan to blow cooling air over the exhaust system when a temperature switch said things were getting too hot. The turbo made the Pulsar a lot hotter. Horsepower went to 100 bhp, a 43-percent increase, and max torque increased to 112 ft-lb at 3,200 rpm.

Nissan rejected a full "European suspension" as too harsh, but the U.S. market Turbo did get as standard equipment: stiffer springs and firmer shocks, and 5.0x13-in cast-aluminum wheels with 175/70SR-13 Toyo radials, which critics faulted for a lack of grip.

When the 1983 Pulsar NX Turbo arrived at Nissan dealers in June 1983, a three-speed automatic was the only transmission available. Nissan had experienced problems with low-rpm vibrations with the five-speed prototypes, so it promised the manual for later. Nissan in-

Except for a lack of tape stripes, this 1985 Pulsar NX is identical to its 1983 predecessors. *Nissan*

stalled a tall 3.17:1 final drive ratio with the turbo engine (instead of naturally aspirated engine's 3.48:1 or the 3.55:1 slated for the turbo stick), and then a low-stall-speed torque converter. As a result, the NX Turbo was slow off the mark, at least until boost came on, shooting the NX forward like it had been given a hotfoot. The NX Turbo was hard to drive, or unpleasant, or both, depending on your point of view.

Whether it was the unconventional bodywork, the underachieving suspension, or the lack of a manual transmission, the Pulsar Turbo failed to set the performance crowd on its collective ear (nor did it help that road tests didn't hit the newsstands until almost the 1984 model year), while other buyers were content with the base engine. The $8,349 Turbo accounted for only two percent of the Pulsar's 1983 annual sales.

1984-1986

Missing in action on Nissan's 1984 roster, the NX Turbo didn't stay long on Nissan's platter. The base Pulsar NX, however, remained popular, its sales even climbing as years went on. Priced at $8,349 in 1986, the Pulsar NX, per a Nissan spokesperson at the time, appealed to a person who wanted a real sports car but couldn't afford one (thereby confessing that a Pulsar wasn't a real sports car?). Almost two-thirds of the nearly 50,000 Pulsars sold in 1985 (up one-third from the previous year) were bought by single young females. Whether Nissan targeted this market or the market just found the car probably doesn't matter except to some lucky product planners at Nissan.

What to Look For

If the Pulsar NX was the "secretary's car" when it was new, an eyeball survey indicates that the Pulsar is still popular with young drivers, especially as a "first car" purchase. Check for a school parking permit. Pulsars have survived in surprising numbers, perhaps because they don't inspire long-distance driving but also because the naturally aspirated engines are anything but overstressed. Rusty cars don't seem to be over represented but young impecunious drivers mean the mechanicals may be neglected, and typical outdoor storage means environmental wear. The Turbo model didn't have a water-cooled turbocharger bearing, so the unit may need to be replaced if it hasn't already been, and it wouldn't make sense to put in a new turbo without at least a partial overhaul. Take care. It could be a financial black hole or, worse, The Project That Was Never Finished.

Don't ever expect huge leaps in value with the NX, even the Turbo, because despite their overall sporty nature, they're (a) just too common, and (b) there's no particular reason—technology, styling, performance, or competition history—to collect them. Someday they'll make great special-interest cars—especially the rare Turbo—but not yet.

1983 Nissan Pulsar NX
Base Price: $7,399

General

Layout	fwd
Curb weight (lb)	2,050
Wheelbase (in)	95.1
Track, front/rear (in)	54.9/54.1
Overall length (in)	162.4
Width (in)	63.7
Height (in)	54.1
Ground clearance (in)	5.9
Seats	5
Trunk space (cu-ft)	14.0 + 10.5
Fuel tank capacity (gal)	13.2

Chassis and Suspension

Frame type	Unit steel
Front suspension	MacPherson struts, lower A-arms, coil springs, anti-roll bar
Rear suspension	Semi-trailing arms, coil springs, tube shocks
Steering type	Rack & pinion, power assist
Steering ratio	18.5:1
Tire size	175/70SR-13
Wheel size (in)/type	13x5/steel

Brakes (vacuum assist)

Front type	Disc
Size (in)	9.4
Rear type	Drum
Size (in)	7.1x1.4

Engine

Type	SOHC Inline 4
Bore & stroke (mm)	76.0 x 88.0
Displacement (cc)	1,597
Compression ratio	9.4:1
BHP @ rpm	69 @ 5,200
Torque @ rpm (ft-lb)	92 @ 3,200
Fuel supply	Hitachi 2V
Fuel required	Regular

Drivetrain

Transmission speeds/types	5/man and 3/auto
Ratios 5-speed (:1)	
1st	3.33
2nd	1.96
3rd	1.29
4th	0.90
5th	0.73
Final drive ratio	3.65

Performance (Source: Road & Track 1/83)

0-60 mph (sec)	12.8
1/4-mile (sec @ mph)	18.8 @ 71.0
Top speed (mph)	101
Fuel economy (mpg)	26.5 observed

separate box

1983 Nissan Pulsar NX Turbo
(same as above except as noted)
Base Price: $8,349

General

Curb weight (lb)	2,130
Track, front/rear (in)	54.9/54.5

Chassis and Suspension

Steering type	Rack & pinion
Wheel type	Cast aluminum

Brakes

Rear type	Drum
Size (in)	8.0x1.4

Engine

Type	OHC turbo Inline 4
Bore & stroke (mm)	76.0 x 82.0
Displacement (cc)	1,488
Compression ratio	7.8:1
BHP @ rpm	100 @ 5,200
Torque @ rpm (ft-lb)	112 @ 3,200
Fuel supply	JECS/Bosch L-Jetronic

Drivetrain

Transmission speeds/types	3/automatic
Ratios 3-speed automatic (:1)	
1st	2.83
2nd	1.54
3rd	1.00
Final drive ratio	3.17

Performance (Source: Car and Driver, 8/83)

0-60 mph (sec)	9.7
1/4-mile (sec @ mph)	17.3 @ 81
Top speed (mph)	109
Fuel economy (mpg)	24 observed

Chapter 18

★	**1987-1990 Pulsar NX**
★1/2	**1987-1988 Pulsar NX SportBak**

1987-1990 Nissan Pulsar NX

Longer, lower, wider: The 1987 Pulsar NX had it over its predecessor, and not only in size. The new Pulsar had style, the old Pulsar had . . . well, it was unique.

Nevertheless, the generations shared major design themes, including the wedgy profile with a notchback and pop-up headlamps. Like its predecessor's, the new Pulsar's windshield was steeply raked while its rear window was relatively vertical. And of course, the new Pulsar had a hip, youthful look about it, just like the old one.

The new Pulsar's shape was the product of the Nissan Design International (in San Diego), and while one might quibble about this detail or that—such as the silver-only hatchback or the taillamps being a series of diagonal slashes—there's no argument that the new NX was

Although still wedge-like overall, the styling of the 1987 Pulsar NX was as sophisticated as the earlier models' had been odd. *Nissan*

more chic than the prior Pulsar, and it had suavity its peculiar little predecessor could never muster.

Under the flash, the Pulsar was essentially new, with different chassis, suspension, and engines. Two powerplants were offered, with the base XE version powered by a 1.6-liter SOHC throttle-body injected four (E16I) that meted out 70 hp at 5,000 rpm. The optional SE got a similar displacement 1.6-liter four, but this one with double overhead cams, four valves per cylinder, and port-fuel injection. The difference was 43 hp. The Pulsar NX SE was rated at 113 bhp.

The 16-valve engine had a split-plenum intake manifold. During low-rpm, low-load conditions, intake air comes from only part of the plenum, with high gas velocities aiding low-speed torque. At 3,600 rpm an electronically controlled butterfly valve opened to allow greater airflow for maximum power. Other leading edge technobits included hot-wire airflow measurement and distributor-less ignition, all controlled by a single engine-control computer. The engine produced its peak 113 bhp at 6,400 rpm, and though the torque topped out at 99 ft-lb at a rather elevated 4,800 rpm, the engine wasn't peaky, making 90 ft-lb of torque at 2,400 rpm.

The SE was available only with a five-speed manual transmission, limiting power to those with an educated left foot, although a three-speed automatic was available on the XE. The engine, per usual small car practice, was mounted transversely and drove the front wheels.

The suspension used familiar MacPherson struts in front but added struts at the rear as well. The rear strut setup had dual lower lateral links and an anti-roll bar acting as trailing links. The XE got 185/70R-13 tires on 5.0-in rims, while the SE's alloys were 6.0 in wide and wrapped with 195/60HR-14 Bridgestone Potenza radials, a major step up from the earlier car. Brakes were disc/drum and met with no complaints.

The 1987 Pulsar NX was heavier than the previous model, much of the weight attributed to stronger structure needed for a mix-and-match roof. While a T-bar roof was standard on all versions of the NX, the hatchback was removable (with a couple of tools, two people,

The original concept of the 1987 Pulsar NX generation had been modular styling. In 1988 the Sportbak option let buyers choose a mini-wagon-back that interchanged with the standard hatchback. *Nissan*

The Sportbak opened in clamshell fashion, and had a high lift-over.

and 15 minutes) to make a mini-pickup truck of sorts. The triangular rear side glass stayed in place, but the rest lifted off. Nissan also promised a station wagon-type roof for a later introduction, while test marketing it in three cities in 1987.

Even Nissan representatives were loath to call the interior 2+2; a better description was a two-seater with a rear seatback that folded up for (very) occasional use as a seat. The interior styling, however, showed that Nissan was getting very good at it, the dash contoured into door panels with neat fabric inserts. Instead of the usual turn signal and et cetera stalks, Nissan went with a late-Eighties trend and mounted these controls all on pods raised out of the dash, supposedly a fingertip's reach away from the steering wheel. It looked like a good idea but would be a rather short-lived fad. Gauges were limited to tach and speedometer, temperature and fuel, with everything else consisting of warning lights.

It should have been a hit with enthusiasts as well as those who just wanted to look hip, but the curb weight of 2,620 lb was too much for 1.6 liters of even the latest tech to tote with alacrity. Even for the SE, zero-to-sixty took over 10 sec. But with prices starting as high as $12,000 to $13,000 (for admittedly well-equipped models), the Pulsar sold well. While 1986 calendar year sales of the Pulsar totaled 47,350, the following year's totaled 62,175. Nissan had a hit.

The SE's DOHC 1.6-liter four produced 61 percent more horsepower than the base XE engine.

1988

The second year after a model introduction is usually a "carry-over" year. The car is still new, sales are still going well. Why waste improvements before you need them to do a little image polishing?

But competition is a great motivator. Nissan quickly slipped a bigger motor into the SE, bringing the 1.6 out to 1.8 liters and the horsepower rating up to 125 bhp. It was a not unwelcome boost. Another mechanical change was the addition of a four-speed automatic transmission to the SE's short option list.

Visibly, the hatchback was painted in body color for 1988, making the Pulsar look less the modular car Nissan hoped people would see it as. The modular look was enhanced by the nationwide availability of the SportBak model. The SportBak was the promised sport wagon model, though instead of a true wagon, it was a hatchback fitted with a squared-off hatch. Although widely hailed as an innovation, the clamshell lid had all the limitations of a fastback design with all the disadvantages of a wagon. The SportBak lid also was available only in charcoal silver, that being the only color in which the sole supplier could produce the fiberglass cover. Dealers began repainting the covers to move SportBaks off their lots, and although the SportBak sold well in California, the rest of the country was apathetic.

Nissan canceled the program within a year. Too bad. The NX SE SportBak at least put some variety on the road.

Nissan had expected demand for the Pulsar NX to be mostly for the smaller-engined XE, but the SE proved so popular that its production mix for 1988 was changed to 40 percent. Prices for 1988 were $11,749 for the base XE, $12,999 for the Sportbak XE. The hatchback SE listed for $12,999, the Sportbak SE for $13,549. Pulsar sales for calendar year 1988 totaled 44,317.

1989

There were no significant changes to the Pulsar line for 1989 except that the Sportbak was gone. Base prices were up to $11,999 for the XE and $12,999 for the SE. Sales were 23,909 units for the calendar year. The SE was dropped for 1990, leaving the XE at $12,249 for the final model year. Only 3,839 were sold in calendar 1990, a final 33 in 1991 as inventory was cleaned up and a replacement sports coupe threw its hat—or perhaps sunroof—into the ring.

What to Look For

The 1987-1990 Pulsar is a "used car" that won't have even special-interest status for a long time. An exception, however, is SportBak-equipped examples. These mini-wagons, if not car show material yet, will turn heads everywhere due to their scarcity. As a commuter with a distinctive look, it's hard to beat. The engine is new enough that you won't have to deal with a carburetor, and of course ignition has been computerized for a long time. The choice is the bigger-engined SE, especially 1988 or later. You get a lot more power in exchange for a marginally higher price and slightly less fuel economy.

1987 Pulsar NS SE
Base Price: $11,799

General	
Layout	fwd
Curb weight (lb)	2,600
Wheelbase (in)	95.7
Track, front/rear (in)	56.7/56.7
Overall length (in)	166.5
Width (in)	65.7
Height (in)	50.8
Ground clearance (in)	5.8
Seats	4
Trunk space (cu-ft)	9.0 + 6.9
Fuel tank capacity (gal)	13.2
Chassis and Suspension	
Frame type	Unit steel
Front suspension	MacPherson struts, lower A-arms, anti-roll bar, coil springs
Rear suspension	Chapman struts, dual lower lateral links, anti-roll bar acting as trailing links, coil springs
Steering type	Rack & pinion, power assist
Steering ratio	n.a.
Tire size	195/60R-14
Wheel size (in)/type	14x6/alloy
Brakes (vacuum assisted)	
Front type	Vented disc
Size (in)	9.8 diameter
Rear type	Drums
Size (in)	7.1x1.4
Engine	
Type	DOHC 16-valve Inline 4
Bore & stroke (mm)	78.0 x 83.6
Displacement (cc)	1,597
Compression ratio	9.0:1
BHP @ rpm	113 @ 6,400
Torque @ rpm (ft-lb)	99 @ 4,800
Fuel supply	Bosch L-Jetronic
Drivetrain	
Transmission speeds/types	5/manual
Ratios 5-speed (:1)	
1st	3.33
2nd	1.96
3rd	1.29
4th	0.98
5th	0.81
Final drive ratio	4.47
Performance (Source: Road & Track, 1/87)	
0-60 mph (sec)	10.3
1/4-mile (sec @ mph)	17.8 @ 79.5
Top speed (mph)	115 estimated
Fuel economy (mpg)	24.0 observed

Chapter 19

1/2	1991-1993 Pulsar NX 1600
★	1991-1993 Pulsar NX 2000

1991-1993 Nissan NX 1600 and NX 2000

Hold on to your socks. Just as the 1987 Pulsar NX had been a revolutionary, not evolutionary, advance over its predecessor, the 1991 Nissan NX 1600 and NX 2000 made the second-generation Pulsars look—and feel—tame. The 2.0-liter's engine zinged to a 7,500-rpm redline while making power-per-cubic-inch that was once limited to turbo cars of the same displacement. Handling was can't-tell-if-it's-front-or-rear-drive, and styling was—if not universally loved—distinctive and anything but boring.

The NX 2000 had unusual recessed ovoid headlamps. The model shown here is a 1992 2000. *Nissan*

The 1991 NX styling was, like its predecessor's, the product of San Diego-based Nissan Design International. It was intended as a radical break from tradition. As NDI Gerald Hershberg vice-president said, "What song should we be singing, and who are we singing it for?"

Possibly Mork from Ork. The profile was so egg-shaped it could have been sold by the dozen. Some thought that cute, others thought it weird. The headlamps were recessed but also ovalish. Only the rear end strayed from the theme—or maybe not—with a ducktail.

The shape alone wouldn't sell these cars, however. While the "fun-to-drive" NX 1600's 16-valve 1.6-liter four produced a respectable 110 bhp (a number that not too long before would have been outstanding), the sporty NX 2000 had 140 eager horsepower under its hood. The SR20DE engine was the same four used in the Nissan Sentra SE-R and the upscale Infiniti G20, and while earlier Nissan engines had been criticized as coarse in the upper rpm ranges (which Nissan engineers had excused as a way to keep drivers from overrevving), the new mill was a zinger. Redline on the 9,000-rpm tachometer was 7,500 rpm, and though the power peak was 6,400 rpm, there was useful power to be had between there and max rpm.

As in the other Nissan models, the 1,998 cc engine had two silent chain-driven overhead camshafts and four valves per cylinder in pent-roof combustion chambers with centrally-mounted, 60,000-mile platinum-tipped spark plugs. Porting in the aluminum head was, per Nissan, "high aerodynamic." Sequential multi-port fuel injection aided not only power but emissions and fuel economy. Steel cylinder liners were cast in the high-strength aluminum block. The engine also had a forged crankshaft and full bearing beam main supports. For an engine for economy (Sentra) and an economy sports car (NX), the 2.0-liter four was packed with technology.

Per common practice for front-drive cars, the engines were mounted transversely and either could be bought with the standard five-speed manual or four-speed, lock-up torque converter automatic. Used with the 2.0-liter engine only, a viscous limited-slip differential on the front axle was an uncommon touch and an aid to getting the power to the road.

The NX 2000 differed from the NX 1600 by the addition of fog lamps, lower body cladding, a small rear spoiler, and alloy wheels. This is a 1993 model. *Nissan*

MacPherson struts constituted the front half of the fully independent suspension, while at the rear, there were struts with two lateral and one trailing links. Anti-roll bars were used front and rear as well. It was like the SE-R, but with firmer springs and anti-roll bars. The NX 1600 came with 13x5-in styled steel wheels with full wheel covers (why mess with the latter when you have the former?) and 175/70R13 radials. The NX 2000 went up a size or two, however, putting 195/55R14 Bridgestone Potenza RE71 on 14x6-in seven-spoked alloy wheels.

Brakes on the NX 1600 were a conventional disk/drum arrangement, but the NX 2000 had discs all around, vented in front and all four bigger than the 1600's. ABS was optional on five-speed NX 2000s. Power-assisted rack-and-pinion steering was standard.

Every NX came with power door mirrors and a tilt steering wheel, and the NX 2000 adds a front air dam with integrated fog lamps, a rear spoiler, and an AM/FM/cassette audio system.

Officially the '91 NXs were two-plus-twos, but don't try to convince a pair of adults who spent any time in the back. Leg room was limited and the roofline may be the source of the term "headbanger." Sitting up front gave a different perspective in attitude and accommodations. Seats were bolstered adequately for autocross use. The steering wheel, with an airbag (no mouse belts!) in the middle, had four

The NX 2000 was successful in SCCA Showroom Stock B racing. David Daughtery won the 1993 national championship. *BF Goodrich*

spokes, the top two at nine and three o'clock. The dash was simple and uncluttered, and though the only accessory gauges were fuel and engine temperature, the 9,000-rpm tachometer and rather optimistic 150-mph speedometer were neat and businesslike.

Getting down to business, testers were enthralled by the NX 2000. *Car and Driver* called it "maybe the world's quickest ovoid," and *Automobile* put the 2000 on its "1992 All Stars" list. The 140 horses and a curb weight of 2,461 lb (2000 5-speed) made it a lively performer and a class bully, even if no stoplight threat to Z28s and such.

Handling was a long suit for the NX 2000, however. A lower center of gravity with shorter tires gave the 2000 a leg up on the SE-R. Said *Road & Track*, "Throw the NX 2000 into a corner at speed and there is absolutely no drama. The body leans a bit, but overall grip of the Bridgestone Potenza RE71s is excellent, and the rear end follows obediently. Lift off the throttle and the tail will swing out to help tighten the cornering line." Enough said.

The only complaints involved a slightly stiff suspension for bumpy roads and apparently slow cycling of the two-circuit ABS that upset balance under racetrack braking. Neither, however, should have been problems for the target market.

Base prices were a reasonable $11,090 for the 1600 and $12,970 for the 2000. Limited availability hurt first-year sales, essentially with only California being served. A mere 5,562 units were sold for the model year. Sixty percent were 2000s, of which slightly more than half had ABS. One-third had automatics.

1992

The NX 1600 and 2000 became available nationwide beginning with the 1992 model year, a factor that made the model significantly more desirable to those living outside of California. Also helping in the desirability sweepstakes were two new options, a new option group and an optional T-bar roof with glass panels. The roof, optional on the 1600 and 2000, wasn't available on any other car in the NX's class. The Power Package, limited to the NX 2000, included cruise control and power windows and door locks, features surprisingly not available on the 1991 NX.

Base prices moved to $11,470 and $13,680 for the 1600 and 2000, respectively. But despite nationwide availability, only 9,389 NX 1600 and 2000s were sold in the '92 model year. Seventy-two percent had the new T-roof option, a quarter had the Power Package, and 58 percent were NX 2000s.

1993

No mechanical changes were made for 1993, though the standard equipment list for the NX 2000 grew by the addition of a four-speaker AM/FM/cassette audio system. Prices went up to $11,820 for the NX 1600 and $14,950 for the NX 2000, and sales slipped again to 7,329 for the model year. But 96 percent imported by Nissan were 1600s. It would be the final year for the NX series, which surely was a disappointment for Nissan, which had seen sales of the Pulsar run steadily in the 40,000 range. But whether it was the unconventional styling or the flood of sport coupes that hit the market in the early Nineties, the NX series—the NX 2000 in particular—remains an excellent choice for enthusiasts, just a little more rare.

What to Look For

The NX 2000 carried a premium over the NX 1600 when it was new and it can be expected that the difference will continue until the cars approach 10 years old. Of course, there's a performance and equipment edge too, so a premium for the 2000 will be well rewarded. The NX 2000 also has a performance edge over the mechanically similar Sentra SE-R, but the price will be higher and the back seat smaller—though the NX 2000 has more cargo room, especially with the back seat folded. But then again, do you like the looks of the NX? Decisions, decisions.

1991 Nissan NX 2000
Base Price: $16,365

General	
Layout	fwd
Curb weight (lb)	2,668
Wheelbase (in)	95.7
Track, front/rear (in)	56.9/56.3
Overall length (in)	162.4
Width (in)	66.1
Height (in)	50.9
Ground clearance (in)	5.2
Seats	4
Trunk space (cu-ft)	14.5+6.3
Fuel tank capacity (gal)	13.2
Chassis and Suspension	
Frame type	Unit steel
Front suspension	MacPherson struts, lower L-arms, coil springs, anti-roll bar
Rear suspension	Chapman struts, dual lower lat eral links, trailing links, coil springs, anti-roll bar
Steering type	Rack & pinion, power assist
Steering ratio	16.5:1
Tire size	P195/55R-14
Wheel size (in)/type	14x6/cast alloy
Brakes	
Front type	Vented disc
Size (in)	10.0 diameter
Rear type	Vented disc
Size (in)	9.1 diameter
Engine	
Type	DOHC 16-valve Inline 4
Bore & stroke (mm)	86.0 x 86.0
Displacement (cc)	1,998
Compression ratio	9.5:1
BHP @ rpm	140 @ 6,400
Torque @ rpm (ft-lb)	130 @ 4,800
Fuel supply	Nissan ECCS with port injection
Drivetrain	
Transmission speeds/types	5/man and 4/auto
Ratios 5-speed (:1)	
1st	3.06
2nd	1.83
3rd	1.29
4th	0.98
5th	0.76
Final drive ratio	4.18
Performance (Source: Road & Track, 2/91)	
0-60 mph (sec)	8.1
1/4-mile (sec @ mph)	16.6 @ 86.0
Top speed (mph)	126 (Car and Driver, 4/91)
Fuel economy (mpg)	30.0 observed

Chapter 20

1991-1994 Nissan Sentra SE-R

The Sentra was the final component to a product lineup that Nissan had completely replaced in the preceding three years, and it was remarkably refined. It was a safe bet to recommend to anyone who wanted an economy car, from the church-mouse-cheap E, to the Mr.-Trump-your-subcompact-is-ready GXE.

Like those magnetic kissing Dutch dolls, enthusiasts snapped their attention to a particular model off to the side of the spare-to-spiffy equipment hierarchy. This was the Sentra SE-R. Touted by Nissan as the car for "enthusiast buyers who want performance without a high price tag," the SE-R had the hot

Fog lamps and an add-on chin spoiler identified a 1991 Sentra SE-R from the front. *Nissan*

Although the rear spoiler was available on other Sentra models, the small SE-R badge was the sure tip-off from the rear about exactly which model you're following. *Nissan*

new all-aluminum 2.0-liter DOHC four-cylinder engine (also used in the NX 2000) in place of the Sentra's usual 100-bhp 1.6-liter four. The 16-valve 140-bhp engine was available only with a five-speed manual transmission (with different ratios than the gearboxes of lesser Sentras), and in the two-door body style only. Like the NX with the big engine, the SE-R came with power rack-and-pinion steering, Nissan's front-axle viscous limited-slip differential, and power four-wheel disc brakes standard; ABS was optional. Of course, like other Sentras, the SE-R had fully independent strut-type suspension, although with "sport" tuning.

In addition to the usual trunk badging, the SE-R could be identified by its own 14-in five-"aero-spoke" alloy wheels, 1/2-in wider than other Sentras' wheels. Tires on the SE-R were bigger as well, at 185/60R14. The SE-R shared body-color side moldings with the GXE, and a rear deck spoiler with the SE. Inside, the SE-R had well-bolstered "sport" bucket seats and all the usual options except power windows. With no airbags, the SE-R, like other two-door Sentras, had annoying door-mounted passive shoulder belts.

The specifications should have been enough to make the impecunious car enthusiast sweat, as should the nimble handling and the 7.6-sec 0-60-mph performance. If not, consider what it would make the competition do.

1992-1994

The Sentra SE-R was essentially unchanged for 1992, but for 1993 an automatic transmission became optional on the SE-R. As part of a product freshening for the 1993 Sentra,

For 1993 the SE-R received new headlights, taillights, and an integrated front spoiler. *Nissan*

all Sentras got a new grille, headlamps, and taillamps, and the SE, GXE, and SE-R had new "rear finisher panels." The SE-R shared a "refined front fascia" with an integrated air dam with the SE. The interior was changed by a new Maxima-like dashboard with better instrumentation and controls. A driver's side air bag was optional, but the passive belts nevertheless remained.

Changes to the 1994 SE-R were limited to the use of non-CFC refrigerant in the optional air conditioning and two new exterior colors.

The SE-R was a kind of "designer" performance car, one that was loved by the press and respected by enthusiasts. But its quart engine in a pint body wasn't enough to draw in the masses, most of whom apparently put style over substance. With an abundance of sporty coupes on the market, the plain-but-potent SE-R didn't get as much attention as Nissan had wished. The base price should have been attractive: $11,370 in 1991 (but remember, no A/C option) to $14,249 in 1994. But even with total Sentra model year sales ranging from 145,000 in 1991 to 172,129 in 1994, the SE-R represented only five percent of 1994 model year North American-built Nissans. To quote *Car and Driver*, "That terrific hot-rod bargain was a marketing flop."

What to Look For

Just because Nissan didn't sell very many in 1991 through 1994 doesn't mean the car won't be a good deal on the used car market. Don't expect anyone to be napping on the SE-R, however, as the Sentra line overall has retained value well. And if you have thoughts of keeping an SE-R long enough to use as some sort of investment, note that owning a Datsun 510 hasn't made anyone rich, and that car has a racing heritage the SE-R can only dream about.

1991 Nissan Sentra SE-R
Base Price: $11,370

General	
Layout	fwd
Curb weight (lb)	2,600
Wheelbase (in)	95.7
Track, front/rear (in)	56.9/56.3
Overall length (in)	170.3
Width (in)	65.6
Height (in)	53.9
Ground clearance (in)	6.0
Seats	5
Trunk space (cu-ft)	16.0
Fuel tank capacity (gal)	13.2

Chassis and Suspension	
Frame type	Unit steel
Front suspension	MacPherson struts, lower A-arms, coil springs, anti-roll bar
Rear suspension	Chapman struts, dual lower later links, trailing links, coil springs, anti-roll bar
Steering type	Rack & pinion, power assist
Steering ratio	15.2:1
Tire size	P185/60R-14
Wheel size (in)/type	14x5 1/2/cast alloy

Brakes (vacuum assist, ABS optional)	
Front type	Vented disc
Size (in)	9.8 diameter
Rear type	Disc
Size (in)	9.1

Engine	
Type	DOHC 16-valve Inline 4
Bore & stroke (mm)	86.0 x 86.0
Displacement (cc)	1,998
Compression ratio	9.5:1
BHP @ rpm	140 @ 6,400
Torque @ rpm (ft-lb)	132 @ 4,800
Fuel supply	Nissan ECCS with port injection

Drivetrain	
Transmission speeds/types	5/manual[1]
Ratios 5-speed manual (:1)	
1st	3.06
2nd	1.83
3rd	1.29
4th	0.98
5th	0.76
Final drive ratio	4.18

Performance (Source: Road & Track, 7/91)	
0-60 mph (sec)	8.1
1/4-mile (sec @ mph)	16.2 @ 87.0
Top speed (mph)	125 estimated
Fuel economy (mpg)	26.0 observed

[1] Four-speed automatic optional in 1993

1995-1996 Nissan 200SX

The scene: An upper floor of an office building in Yokohama, Japan.

The players: Two Nissan executives.

The time: Early 1992, after the 1991 sales data is in.

Nissan exec #1: The NX 2000 isn't going over too well in America.

Nissan exec #2: The Sentra is doing well, but the SE-R's sales aren't too great.

Exec #1: They say it might be that the styling of the NX is too *avant- garde*.

Exec #2: Well, they love the performance of the SE-R, but they say it looks too ordinary.

#1: Hey, I have an idea: If the NX is too outlandish, and the Sentra SE-R too plain. . . .

#2: Why don't we combine the two into one car? We can keep the neat engine and have stylist cook up something in between the NX 2000 and the Sentra SE-R.

#1: And you know, we can make the suspension a little less expensive while we're at it.

#2: Not to mention adding de-contented versions.

It probably didn't happen that way, but the results look like it could have. The car in question is the 200SX. What it is not is a replacement

The 1995 200SX SE-R combined an old name with the performance suffix from the discontinued Sentra SE-R. *Nissan*

The 200SX SE-R came standard with alloy wheels and the obligatory rear spoiler. *Nissan*

for the 240SX—a new version of which was also introduced in 1995—despite that model's origin as a 200SX. It's not a Sentra, the '95 version of which was introduced in early 1995, but it is Sentra-based. It's not an NX, its styling being much more conservative and Nissan having no particular reason to retain that name.

The 200SX gave Nissan a lot more mileage from the basic Sentra elements, the economy sedan sharing its chassis and drivetrain with the sportier coupe. That meant the new 200SX received all the benefits of the work that went into making the chassis of the Sentra more rigid, as well as the new corporate rear suspension used on the Sentra and the Maxima.

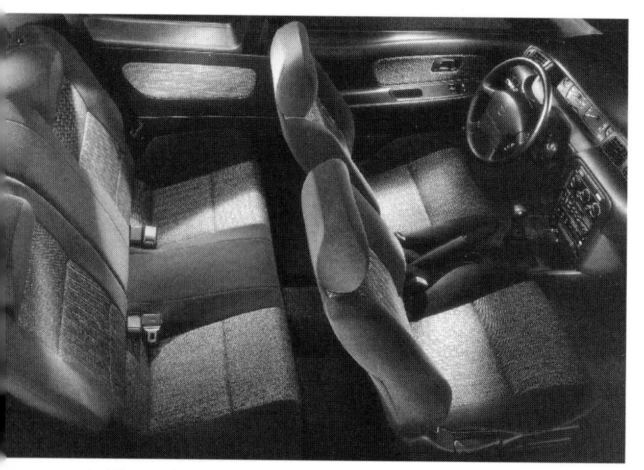
Though best suited for two, the 200SX SE-R had seating for up to five. *Nissan*

The 200SX lineup included base and SE models powered by the 1.6-liter DOHC four of the earlier base Sentra and NX 1600. The engine had been reworked to put out 115 bhp, five more than before, and get better fuel mileage. The 200SX SE-R, on the other hand, gets its motivation from the 2.0-liter DOHC 16-valve engine of the previous Sentra SE-R and NX 2000, a little engine that motivates like Dale Carnegie. As before, the engine was rated at 140 bhp and 132 ft-lb of torque, though for the sake of emissions, Nissan lopped 400 rpm off the redline.

The 2.0-liter could be mated to an optional four-speed automatic as well as the standard five-speed manual. The front-axle viscous limited-slip differential was retained, as was power rack-and-pinion steering. The base 200SX had disc/drum brakes, while the SE-R had four-wheel discs with ABS optional. (The SE had rear drums standard, and rear discs with the optional ABS)

Wheel size ranged from 5.0x13-in steel wheels on the base car, to 5.5x14-in alloys on the SE, to 5.5x15-in alloys on the SE-R. Tires, respectively, went from 175/70R13, to 175/65R14 and 195/55R15.

The front suspension consisted of the ubiquitous MacPherson struts, but at the rear was an ingenious assemblage called a Multi-Link Beam. It combined two trailing arms and a transverse beam axle with a lateral link and control rod, a sort of specially mounted Panhard rod. Nissan claimed improved ride and handling, lower weight and decreased road noise, along with more room in the trunk and passenger compartment. The trade off would be decreased ability to absorb one-wheel bumps and, to some testers, a decrease in cornering turn-in.

The interior was neat and even luxurious, the SE and SE-R equipped with "sport" front bucket seats, well bolstered but not intrusively so. Instrumentation, in the four-gauge mode (tach, speedo, temp, and fuel), was handsome, and the controls were easy to use. The big difference—at least from the NX—was in the back. Despite the sloped backlight of the 200SX, headroom was adequate for adults (though down 1/2 in from the old sedan-roofed Sentra SE-R). Legroom made the back seat—unlike that of the NX—a functional piece of furniture, but the additional 4 in of wheelbase the 200SX had over the old Sentra yielded only 1/2 in more length for the legs. It still was no skybox back there, but with the split fold-down rear seatbacks, skis or other long objects could be transported inside despite 200SX's conventional trunk.

Critics being critics, the new exterior styling took its lumps as being too "average"—perhaps from splitting the difference between the old Sentra and NX?—but the 200SX looked good without offending anyone, and had a Cd of 0.33 as well. The SE-R shared the integrated fog lamps and body-color side moldings with the SE, though the side sill extensions were on the SE-R alone. The rear deck spoiler was standard on the SE-R and optional on the SE, though the two models could be easily differentiated by the SE's six-spoke alloy wheels and the five-spoke wheels on the SE-R. And of course, there was the SE-R badge on the latter's trunk.

An advantage the 200SX SE-R brought to the table was price. It was actually less expensive than the earlier Sentra SE-R. But the problem faced by the 1995 200SX would be developing its own personality and getting attention in a crowded small coupe market. It wouldn't be easy.

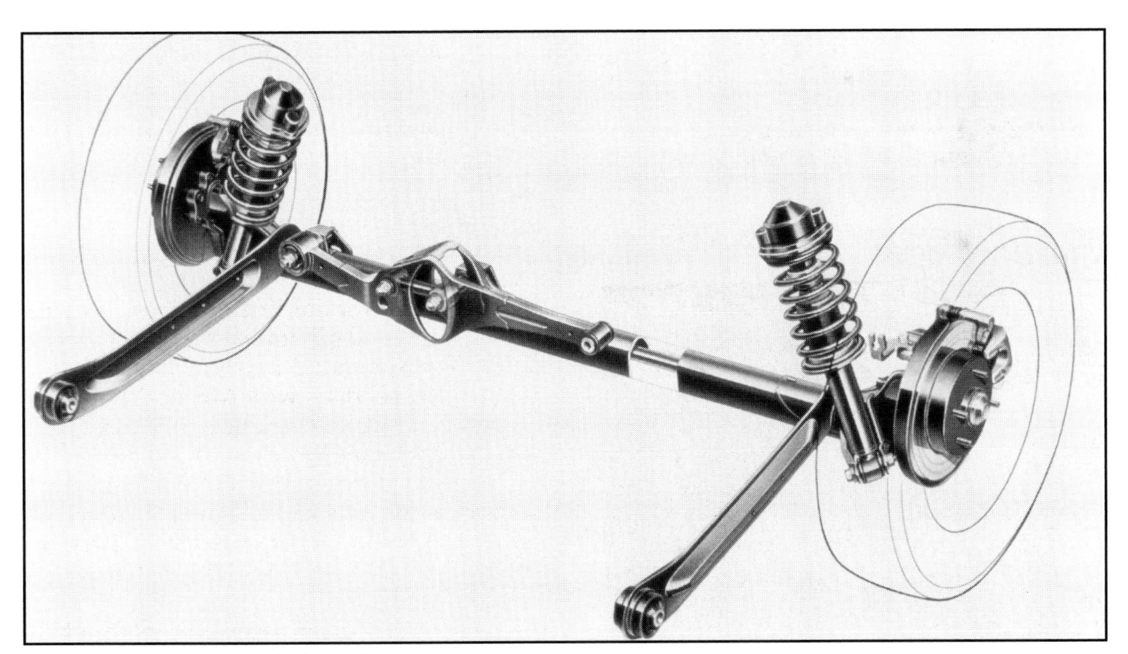

The 200SX abandoned independent rear suspension for a sophisticated multi-link beam rear suspension. *Nissan*

Prices for the 1995 200SX started at $11,999 for the base model, went to $14,269 for the SE, adding another $1,000 for the SE-R. An automatic cost $800 more and anti-lock brakes, not available on the base, added $995 to the SE or SE-R.

1996

As an all-new model in 1995, changes were few for 1996, limited to new body-colored door handles on the SE and SE-R models. Prices for the 1996 200SX were, for the base model, $12,449, rising to $14,869 for the SE, and $16,069 for the SE-R.

What to Look For

Follow the usual used-car caveats when buying a 1995-1996 200SX. Note, too, that slow sales will prompt Nissan to make changes. That won't affect the utility of the 200SX, but may cause it to become outdated sooner.

1995 Nissan 200SX SE-R
Base Price: $15,269

General	
Layout	fwd
Curb weight (lb)	2,588
Wheelbase (in)	99.8
Track, front/rear (in)	57.9/56.5
Overall length (in)	170.1
Width (in)	66.6
Height (in)	54.2
Ground clearance (in)	4.7
Seats	5
Trunk space (cu-ft)	10.0
Fuel tank capacity (gal)	13.2
Chassis and Suspension	
Frame type	Unit steel
Front suspension	MacPherson struts, lower con trol arms, coil springs, anti-roll bar
Rear suspension	Beam axle with two trailing arms, sliding Panhard rod with central link, coil springs, anti-roll bar
Steering type	Rack & pinion, power assist
Steering ratio	n.a.
Tire size	P195/55VR-15
Wheel size (in)/type	15x5.5/cast aluminum
Brakes (vacuum assist, ABS optional)	
Front type	Vented disc
Size (in)	9.9 diameter
Rear type	Disc
Size (in)	9.3 diameter
Engine	
Type	DOHC 16-valve Inline 4
Bore & stroke (mm)	86.0 x 86.0
Displacement (cc)	1,998
Compression ratio	9.5:1
BHP @ rpm	140 @ 6,400
Torque @ rpm (ft-lb)	132 @ 4,800
Fuel supply	Nissan ECCS with port injection
Drivetrain	
Transmission speeds/types	5/man and 4/auto
Ratios 5-speed (:1)	
1st	3.06
2nd	1.83
3rd	1.29
4th	0.98
5th	0.76
Final drive ratio	4.17
Performance (Source: Car and Driver, 2/95)	
0-60 mph (sec)	8.0
1/4-mile (sec @ mph)	16.1 @ 83
Top speed (mph)	109 (governor limited)
Fuel economy (mpg)	26 observed

Recommended Reading

Datsun Z-Cars, Publications International, Skokie, Illinois, 1981.

Fisher, Bill, and Bob Waar, *How to modify Datsun engines and chassis 510, 610, 240Z*, H.P. Books, Tucson AZ, 1973.

Hollander, Michael F., *The Complete Datsun Guide*, Tab Books, Blue Ridge Summit, Pa., 1980.

Hutton, Ray, *Nissan 300ZX*, Motor Racing Publications Limited, Croydon, England, 1990.

Hutton, Ray, *The Z-Series Datsuns Including Nissan 300ZX*, Motor Racing Publications Ltd, Croydon, England, 1985.

Milspaugh, Ben P., *Z car: a legend in its own time*, TAB Books, 1991.

Petersen's Complete Book of Datsun, Petersen Publishing Co., Los Angeles, 1975.

Rae, John B., *Nissan/Datsun: A History of Nissan Motor Corporation in U.S.A. 1960-1990*, McGraw-Hill, New York, 1982.

Ruiz, Marco, *Japanese Car*, Portland House, New York, 1982.

Schrader, Halwart, *The History of Datsun Automobiles*, Schrader & Partner GmbH, Munich, Germany, 1976.

Waar, Bob, *How to Hotrod & Race Your Datsun*, Steve Smith Autosports, Santa Ana CA, 1984 (revised edition).

Index